THE POPPER-CARNAP CONTROVERSY

THE POPPER-CARNAP CONTROVERSY

by

ALEX C. MICHALOS
University of Guelph

MARTINUS NIJHOFF / THE HAGUE / 1971

© *1971 by Martinus Nijhoff, The Hague, Netherlands*
All rights reserved, including the right to translate or to reproduce this book or parts thereof in any form

ISBN 90 247 5127 6

PRINTED IN THE NETHERLANDS

For my brother, Chuck

TABLE OF CONTENTS

Acknowledgements	IX
CHAPTER I: Introduction	1
CHAPTER II: Acceptability and Logical Improbability	3
CHAPTER III: Two Explicanda and Three Arguments	16
CHAPTER IV: Bar-Hillel's "Comments" and Unrestricted Universals	34
CHAPTER V: Instance and Qualified-Instance Confirmation	49
CHAPTER VI: The Singular Predictive Inference	60
CHAPTER VII: Lakatos on Appraisal, Growth and Analytic Guides	70
CHAPTER VIII: Hintikka and Hilpinen on Inductive Generalization	83
CHAPTER IX: Cost-Benefit Versus Expected Utility Acceptance Rules	92
List of Reference	117
Index	123

ACKNOWLEDGEMENTS

Several people have given me advice, criticism and encouragement about various parts of this book as it has developed in different forms the past seven years. I would like to express my gratitude to them and to insist that they should not be held responsible for the finished product. They are Y. Bar-Hillel, M. Brand, S. Bromberger, R. Carnap, C. A. Hooker, D. Hull, S. A. Kleiner, H. Lehman, K. Lehrer, W. Leinfellner, G. Maxwell, A. McLaughlin, H. Mehlberg, W. C. Salmon, T. W. Settle, D. Shapere and J. O. Wisdom.

Parts of the following articles have been reprinted with the permission of the editors and publishers of the journals in which they appeared.

"Two theorems of degree of confirmation", *Ratio*, 7, 1965, pp. 196–198.
"Descriptive completeness and linguistic variance," *Dialogue*, 6, 1967, pp. 224–228.
"Analytic and other 'dumb' guides of life," *Analysis*, 30, 1970, pp. 121–123.
"Cost-benefit versus expected utility acceptance rules," *Theory and Decision*, 1, 1970.
"Rules of acceptance and inductive logic: A critical review," *Philosophy of Science*, to be published.

University of Guelph   Alex C. Michalos
Ontario
August 1970

CHAPTER I

INTRODUCTION

In 1954[1] Karl Popper published an article attempting to show that the identification of the quantitative concept *degree of confirmation* with the quantitative concept *degree of probability* is a serious error. The error was presumably committed by J. M. Keynes, H. Reichenbach and R. Carnap.[2] It was Popper's intention then, to expose the error and to introduce an explicatum for the prescientific concept of degree of confirmation.

A few months later Y. Bar-Hillel published an article attempting to show that no serious error had been committed (particularly by Carnap) and that the problem introduced by Popper was simply a "verbal one."[3]

Popper replied immediately that "Dr. Bar-Hillel forces me [Popper] now to criticize Carnap's theory further," and he [Popper] introduced further objections which, if accepted, destroy Carnap's theory.[4]

About eight years after this exchange took place I was in graduate school at the University of Chicago in search of a topic for a doctoral dissertation. An investigation of the issues involved in this exchange seemed to be ideal for me because I had (and still have) a great admiration for the work of both Carnap and Popper. A thoroughly revised and I hope improved account of that investigation appears in the first five chapters of this book.

Put very briefly, what I found were four main points of contention. First, Popper insisted that Carnap's explicatum[5] of degree of con-

[1] Karl Popper, "Degree of Confirmation," *B.J.P.S.*, V (1954), pp. 143–149.
[2] Popper, "Degree of Confirmation," *loc.cit.*, p. 145, note.
[3] Y. Bar-Hillel, "Comments on 'Degree of Confirmation' by Professor K. R. Popper," *B.J.P.S.*, VI (1955), pp. 155–157.
[4] Karl Popper, "'Content' and 'Degree of Confirmation': A Reply to Dr. Bar-Hillel," *B.J.P.S.*, VI (1955), pp. 157–163.
[5] I am adopting Carnap's terminology. An explication has two elements. The *explicandum* is the prescientific or presystematic concept to be explicated. The *explicatum* is the systematic

firmation was supposed to but could not be an adequate measure of the acceptability of scientific theories. Carnap agreed that it could not be such a measure and insisted that it was not intended to be. Second, Popper tried to demonstrate that Carnap's explicatum could not be an adequate measure of some presumed explicandum that he (Popper) assumed both he and Carnap were attempting to explicate. As it turned out, the two philosophers again evidently had different explicanda in mind. Third, Popper claimed that Carnap's explicatum was unacceptable because it contained an unsatisfactory treatment of unrestricted universal sentences, and this claim was perfectly warranted. Finally, Popper argued that Carnap's explicatum was inconsistent in view of its theorem for singular predictive inferences, but this was a mistake.

Several interesting side-issues arose with these primary problems and several people were drawn into the dispute in one way or another since 1954. In 1968 Imre Lakatos published a long paper reviewing much of the history and many of the issues examined in my dissertation.[6] But Lakatos is more of a historian and popperian than I am, and his essay reveals a number of important points that had slipped by me. Thus, I have included an examination of some of his basic views in chapter VII. Following that there is an analysis of aspects of Jaakko Hintikka's system of inductive logic,[7] in particular, of his theorems and acceptance rules for unrestricted universal sentences. Hintikka's system is an apparently "natural" but nevertheless brilliant step beyond Carnap's, and it deserves much more attention than it has received so far. Finally, I have included a comparison of the Bernoulli-Bayes rule of acceptance with my (by no means entirely novel) rule of cost-benefit dominance. Although I believe that the results of the comparison clearly support my contention that the rule recommended there has great promise, I am fully aware that there is a difference between the promise and the delivery. Some important aspects of that "delivery" shall be made in my *Foundations of Decision Making* (American Elsevier Publishing Co.) which is now in preparation.

or scientific concept. *Explicanda* and *explicata* are simply the plural forms of these terms. See Rudolf Carnap, *The Logical Foundations of Probability* (Chicago: University of Chicago Press, 1950), pp. 3–8.

[6] I. Lakatos, "Changes in the problem of inductive logic," *The problem of Inductive Logic*, ed. I. Lakatos (Amsterdam: North Holland Pub. Co., 1968), pp. 315–417.

[7] Elements of the system have appeared in several papers. See the first note in Chapter VIII.

CHAPTER II

ACCEPTABILITY AND LOGICAL IMPROBABILITY

1. INTRODUCTION

In *The Logic of Scientific Discovery*, Karl Popper introduced and defended the view that the acceptability of a scientific hypothesis is directly proportional to its logical *im*probability.

He assumed that his view was exactly the opposite of Carnap's and he criticized Carnap's explicatum of degree of confirmation for being an inadequate measure of *acceptability*. This is one source of the dispute between Carnap and Popper, and the first I shall investigate.

In section two I shall produce some evidence for Popper's view that Carnap's explicatum has as its explicandum the ordinary or prescientific notion of acceptability (or degree of acceptability). But it will be shown that by 1955 (i.e., by the time Popper began mentioning Carnap in particular in his criticisms of "inductivism") Carnap explicitly denies that his explicandum is acceptability. The first issue of the dispute then, is resolved by showing that Popper is criticizing Carnap for holding a view that Carnap does not in fact hold. Carnap's explicatum must not be expected to measure degree of acceptability or criticized for not measuring degree of acceptability.

The concept of acceptability is as ambiguous and troublesome as probability, confirmation, belief, justice, etc. In this chapter it shall be assigned *two* specific senses, one applicable to hypotheses in virtue of their being worthy of serious investigation and the other applicable in virtue of their having already endured some serious investigation. A hypothesis that is acceptable in the first sense has been called "testworthy" and "bold" by Popper and Lakatos,[1] and one that is acceptable in the second sense has been called "well-tested." For ease

[1] Lakatos, *op. cit.*, p. 380.

of reference, Lakatos refers to acceptability in the "testworthy" sense as 'acceptability$_1$' and acceptability in the "well-tested" sense as 'acceptability$_2$,' and I shall follow this procedure.[2] Carnap's "Rule of Maximizing Estimated Utility" may be used as a measure of acceptability in virtually *any* sense, and I shall criticize it in detail in Chapter IX. Here, after presenting (in section three) Popper's argument in defense of his original explicatum of degree of acceptability$_1$, I consider six plausible objections to his argument, in section four. So far as I know, no one before Popper has presented the argument leading to the *prima facie* incredible conclusion that *im*probable hypotheses are more acceptable$_1$ than probable ones and criticized it in detail. The result of this investigation is a refutation of Popper's argument by my fifth and sixth objections.

2. ACCEPTABILITY AND DEGREE OF CONFIRMATION

For clarity and ease of reference in the discussion that follows, it will be worthwhile to introduce the following abbreviations.

Explicanda

OC_1 means degree of confirmability
OC_2 means degree of confirmation
OA_1 means degree of acceptability$_1$
OA_2 means degree of acceptability$_2$

Explicata

C_1 means degree of confirmability
C_2 means degree of confirmation
A_1 means degree of acceptability$_1$
A_2 means degree of acceptability$_2$

There is no doubt that as far as Popper is concerned there is no important difference between OA_1 and OC_1 on the one hand and OA_2 and OC_2 on the other.[3] But he has also claimed that

'Carnap always intended (as [Popper] did) that his [explicatum] degree of confirmation should serve as a measure of the acceptability of a theory.'[4]

---

[2] He also defines two other senses of acceptability, *ibid.*, pp. 391–399.
[3] See for example, Popper, *The Logic of Scientific Discovery*, pp. 108–109, 270–272 and 394–395.
[4] Popper, "'Content' and 'Degree of Confirmation': A Reply to Dr. Bar-Hillel," *loc.cit.*, p. 162.

What exactly does this mean? Is he saying that Carnap's C_2 was intended as an explicatum of OA_1, OA_2 or both? If the first or the third, then he is charging Carnap with or may himself be confusing the (prior) confirmability and the (posterior) degree of confirmation actually received by a hypothesis. If the second, then he is not charging Carnap with and is not himself confusing prior and posterior appraisals. I do not know what Popper had in mind when he made that claim, but it is worthwhile to try to find some evidence in Carnap's work which would tend to support or undermine it in either of the senses in which it might have been intended. Let us consider the evidence.

In 1936 Carnap wrote

We shall now examine more closely the concept of confirmation. This will require that we *describe* the procedure of scientific testing and that we specify the conditions under which a statement, *as a result of such testing*, is considered as *more or less confirmed*, i.e., scientifically *accepted or rejected*.[5]

(Italics added.)

In this passage Carnap seems to be referring to OA_2 and OC_2. He is going to *"describe* the procedure" and "specify conditions," I gather, that seem to be revealed by the description. There is a remarkable similarity between this passage and many of Popper's. Consider only the following from *The Logic of Scientific Discovery*.

Instead of discussing the 'probability' of a hypothesis we should try to assess what tests, what trials, it has withstood, that is; we should try to assess how far it has been able to prove its fitness to survive *by standing up to tests*. In brief, we should try to assess how far it has been 'corroborated.'[6]

The only dubious aspect of the above quotation is the apparent assumption that *"by standing up to tests"* a hypothesis proves "its fitness to survive" future tests. As Lakatos has insisted, the past performance of a hypothesis is not a quarantee of its present "fitness to survive."[7] In a new footnote (added in 1959) following this passage Popper claims that Carnap was originally only following Popper's usage.

Carnap translated my [Popper's] term 'degree of corroboration' (*'Grad der Bewahrung'*), which I had first introduced into the discussions of the Vienna Circle, as 'degree of confirmation.'[8]

---

[5] Rudolf Carnap, "Truth and Confirmation," *Readings in Philosophical Analysis*, ed. H. Feigl and W. Sellars (New York: Appleton-Century-Crofts, Inc., 1949), p. 124.
[6] Popper, *The Logic of Scientific Discovery*, p. 251.
[7] Lakatos, *op.cit.*, p. 405.
[8] Popper, *loc.cit.*

Carnap's emphasis on testing in the passage above as well as his explicit identification of 'confirmed' and 'scientifically accepted' clearly supports Popper's view that Carnap (at one time) did not distinguish OA_2 from OC_2. Hence, it also supports the view that a measure of OC_2 designed by Carnap *should* also be a measure of OA_2.

The passage from 1936 is not just an isolated case. Consider the following remark from 1945.

> ... we should not ask whether a given statement is true but only whether it has been *confirmed, corroborated, or accepted* by a certain person at a certain time.[9] (Italics added).

And finally, consider two remarks published in 1955.

> Laplace's explanations show clearly that he is mostly concerned, not with actual frequencies, but with methods for *judging the acceptability* of assumptions, *in other words, with inductive probability*.[10] (Italics added.)

> ... inductive probability, as I see it, does not occur *in* scientific statements, concrete or general, but only in judgments *about* such statements; in particular, in judgments about the *strength of support* given by one statement, the evidence, to another, the hypothesis, and *hence* about *the acceptability* of the latter on the basis of the former.[11] (Italics added.)

Clearly then, there is evidence for Popper's view that Carnap's explicatum C_2 was to measure OA_2 and OC_2, because there is evidence that Carnap did not distinguish OA_2 from OC_2. Indeed, in the passages quoted above, he uses 'accepted,' 'confirmed,' 'supported' and 'made probable' as synonyms.

But beginning with *The Logical Foundations of Probability*, one can find fairly conclusive evidence that Carnap's explicatum C_2 was not intended to measure OA_2, but only OC_2. Nevertheless, even in that book we still find evidence for Popper's view. For example, on page 201 we find

> The question whether the premise (evidence) e is known (*well established, highly confirmed, accepted*), is irrelevant...

---

[9] Rudolf Carnap, "The Two Concepts of Probability," *Readings in the Philosophy of Science*, ed. H. Feigl and M. Brodbeck (New York: Appleton-Century-Crofts, Inc., 1953), p. 455. This is a reprint of the article which originally appeared in *Philosophy & Phenomenological Research*, Vol. V (1945).

[10] Rudolf Carnap, "Statistical and Inductive Probability," *The Structure of Scientific Thought*, ed. E. H. Madden (Boston: Houghton Mifflin Co., 1960), p. 272. The article originally appeared as a pamphlet published by the Galois Institute of Mathematics and Art, Brooklyn, N.Y., 1955.

[11] *Ibid.*, p. 274.

But later we find a very strong denial that C_2 can measure OA_2.

> Inductive logic i.e., ['probability logic'] alone does not and cannot determine the best hypothesis on a given evidence, if the best hypothesis means that which good scientists would prefer. This preference is determined by factors of many different kinds, among them logical, methodological, and purely subjective factors.[12]

Here, I take it, we must assume that by "best hypothesis" he means 'most acceptable$_2$ hypothesis.' Even more significant is the fact that he devotes nearly thirty pages of *The Logical Foundations of Probability* to a discussion of rules for "determining decisions" and ends by defining the "Rule of Maximizing the Estimated Utility."[13] Since 1950 (aside from the remarks in "Statistical and Inductive Probability") Carnap has defended this rule as a measure of acceptability in virtually every sense of this term.[14]

What then, shall be our judgement concerning Popper's claim that "Carnap always intended ... [C_2] ... as a measure of the *acceptability* of a theory?" Before 1950 one can find passages in which the synonymity of OA_2 and OC_2 is apparently taken for granted but one can *not* find an explicit denial that OA_2 is synonymous with OC_2. After 1950 one can find passages in which the synonymity of OA_2 and OC_2 is apparently taken for granted but one can *also* find explicit denials of the claim that OA_2 is synonymous with OC_2. After 1955 one can *only* find explicit denials of the claim that OA_2 is synonymous with OC_2 and that Carnap's C_2 might be used as an explicatum of some kind of acceptability.

Therefore our judgment about Popper's claim must be that it is *false* that "Carnap always intended ... [C_2] ... as a measure of the *acceptability* of a theory" in *any* sense of 'acceptability.' At one time Carnap evidently did have this intention, but beginning with his major publication on the subject, *The Logical Foundations of Probability*, he explicitly denies this intention. And that just means that Popper's criticism of Carnap's explicatum C_2 *as an adequate measure of* OA_2[15]

---

[12] Carnap, *The Logical Foundations of Probability*, p. 221.
[13] *Ibid.*, pp. 252–279.
[14] See, for example, "Replies and Systematic Expositions," *The Philosophy of Rudolf Carnap*, ed. P. A. Schilpp ("Library of Living Philosophers," Vol. II; La Salle, Illinois: Open Court, 1963), pp. 972 and 968; and "The Aim of Inductive Logic," *Logic, Methodology and Philosophy of Science*, ed. E. Nagel, P. Suppes and A. Tarski (Stanford, California: Stanford University Press, 1962), p. 304.
[15] For example, in the article cited above in note 2; in the introductory remarks to Appendix *ix in Popper, *The Logic of Scientific Discovery;* and in "Adequacy and Consistency: A Second Reply to Dr. Bar-Hillel," *B.J.P.S.*, VII (1956), p. 249.

8   ACCEPTABILITY AND LOGICAL IMPROBABILITY

is based on a mistaken assumption. Popper assumes, not entirely without justification but nonetheless erroneously, that Carnap's C_2 is supposed to measure OA_2, i.e., that the *explicandum* of Carnap's C_2 is OA_2. Thus, having shown that Carnap's C_2 *is not* intended to measure OA_2, we have shown that Popper's attack is bound to be irrelevant. That is, we have resolved the *first* basic issue of the dispute between Popper and Carnap. Popper claims Carnap's C_2 is inadequate because it fails to measure OA_2; but Carnap's C_2 is not intended to measure OA_2. Thus, C_2 cannot be inadequate for *this* reason.

3. ARGUMENT FROM THE DESCRIPTION OF THE WORLD

Philosophers who use a logical interpretation[16] of the calculus of probability have assigned the highest probability, namely unity, to analytic sentences and the lowest probability, zero, to self-contradictory sentences. If we let 'p(H)' stand for 'the initial probability of the hypothesis[17] H' then $p(H) = 0$ when H is self-contradictory and $p(H) = 1$ when H is analytic.[18] If we let 'p(H, e)' stand for 'the conditional (or relative) probability of the hypothesis H given some (additional) evidence e' then $p(H, e) = 0$ when e contradicts H and $p(H, e) = 1$ when H is entailed by e.[19] If we grant that analytic sentences tell us nothing about the world because they place no restrictions or set no limits on the character of the world then we must grant that the higher the initial logical probability a given hypothesis H has, the less H asserts about the world. Alternatively, considering only the relative probability of H on some evidence e, we may say that the higher the relative probability of H given e, the less H asserts about the world *beyond* what is asserted by e. In general then, we may define the initial empirical content of H 'Ct(H)' and the relative empirical content of H given e 'Ct(H, e)' as follows:

$$\text{Ct(H)} \underset{df}{=} 1 - p(H) \qquad \text{Ct(H, e)} \underset{df}{=} 1 - p(H, e).$$

---

[16] A thorough review of interpretations and theories of measuring probability values may be found in Alex C. Michalos, *Principles of Logic* (Englewood Cliffs, New Jersey: Prentice-Hall, Inc., 1969), Chapter Five.

[17] I am using the term 'hypothesis' in a broad sense to cover ordinary sentences as well as laws and theories.

[18] Strictly speaking, a probability functor requires two arguments. But it simplifies our notation considerably to represent initial probability (probability relative to given initial conditions) as we have it here. This convention is usually followed by Popper too.

[19] When e is false, p(H,e) is, in most systems undefined. Popper has constructed systems in which p(e) has a definite value even when e is false, i.e., when $p(e) = 0$. See Karl Popper, "A Set of Independent Axioms for Probability," *Mind*, XLVII (1938), p. 275.

That is, we use the complement of the probability of H as a definition and a measure of the empirical content of H.[20]

Popper's first and fundamental argument in defense of the view that hypotheses with high (initial *or* relative) probabilities are unacceptable$_1$ runs thus.[21] One of the primary goals of a scientist is the discovery of laws or theories that adequately describe the world in which we live. The elimination of a number of logically possible but erroneous descriptions of "our particular world" is a necessary condition of determining an adequate description of "our particular world." Of all the hypotheses we have, those with the highest (initial or relative) probabilities contribute the least toward the elimination of inadequate but logically possible descriptions of our world. Therefore, of all the hypotheses we have, those with the highest probabilities are the least acceptable$_1$, i.e., the higher the probability of a hypothesis, the more unacceptable$_1$ the hypothesis tends to become.[22] Hereafter, I will refer to this argument as the 'description argument.'[23] If this argument is sound then all those who think that a high logical probability is a sign of acceptability$_1$ are mistaken.[24]

4. SIX OBJECTIONS TO THE DESCRIPTION ARGUMENT

There are at least six fairly plausible objections that one might raise against the preceding argument. I shall present the objections and

---

[20] This particular definition of empirical content is accepted by Popper but, as he points out on p. 270 of Popper, *The Logic of Scientific Discovery*, in 1934 he wavered between using the complement and the reciprocal of probability as definiens. Practically speaking, it makes little difference.

[21] That is, as it is presented in Popper, *The Logic of Scientific Discovery*, pp. 269–270; Popper, "Degree of Confirmation," *loc. cit.*, p. 146; and Karl Popper, *Conjectures and Refutations* (New York: Basic Books, 1963), pp. 286–287.

[22] Popper, *The Logic of Scientific Discovery*, p. 113. It is, I think, worthwhile to quote Popper. "Now theoretical science ... aims at restricting the range of permitted events to a minimum; and, if this can be done at all, to such a degree that any further restriction would lead to an actual empirical falsification of the theory. If we could be successful in obtaining a theory such as this, then this theory would describe 'our particular world' as precisely as a theory can; for it would single out the world of 'our experience' from the class of all logically possible worlds of experience with the greatest precision attainable by theoretical science. All the events or classes of occurrences which we actually encounter and observe, and only these would be characterized as 'permitted'."

[23] On p. 270 of the 1959 edition of *The Logic of Scientific Discovery* Popper has a long footnote that is apparently supposed to summarize the description argument but that actually introduces an entirely different argument involving a confusion between prior corroborability and posterior corroboration and the false assertion that if, say, $p(H) < p(K)$ then it must be the case that $p(H,e) < p(K,e)$ for any evidence e.

[24] It is, in fact, the work of Keynes and Reichenbach that is particularly criticized in *The Logic of Scientific Discovery*, Sections 80 and 83.

suggest the sorts of replies that I think Popper could make. The last two objections are presumably fatal for that argument.

According to the first premise of the description argument, one of the primary goals of a scientist is the discovery of laws or theories that adequately *describe* the world in which we live. But instrumentalists and others have pointed out that scientific laws and theories should not be thought of as descriptions (of regularities or relationships) at all, since most of them are known to be false.[25] In Michael Scriven's words, "Physical laws are not even probably true – they are already known *not* to be true."[26] Rather one should take the view that laws and theories are instruments or tools used for making predictions. They are inference rules, i.e., rules prescribing or permitting certain inferences from certain given statements of fact.[27]

A first objection to the description argument then, might be this: If the instrumentalist view of laws and theories is correct then the first premise of the description argument is false. If laws and theories are *not* descriptions then it is clearly a mistake to assume scientists are interested in discovering laws and theories that *are* descriptions.

Popper's reply to the instrumentalist view is that it fails to take into consideration the implication of the fact that laws and theories *can* be falsified or refuted while instruments can *not* be refuted or falsified.[28] Hence, since laws and theories have a property of descriptive statements (viz., that they can be false or true) but instruments for prediction lack this property, laws and theories must not be *only* instruments for prediction. It follows from this then, that the discovery of adequate descriptions of our world might well be a primary goal of scientists. That is, the first premise of the description argument has not been falsified.

Second, one might point out that a rule such as A: 'Accept$_1$ that hypothesis which has the lowest probability' is self-defeating because one of the goals of scientific research is the discovery of true sentences about the world and according to this rule, we should accept$_1$ sentences that are nearly self-contradictory (i.e., if we are considering initial probability only) or probably false (i.e., if we are considering relative probability).

---

[25] For example, see S. Toulmin, *The Philosophy of Science* (New York: Harper and Brothers, 1960), pp. 53–56 and 100–104.

[26] Michael Scriven, "The Key Property of Physical Laws – Inaccuracy," *Current Issues in the Philosophy of Science*, ed. H. Feigl and G. Maxwell (New York: Holt, Rinehart and Winston 1961), p. 92.

[27] Toulmin, *op. cit.*, pp. 100–104.

[28] Popper, *Conjectures and Refutations*, pp. 111–114.

ACCEPTABILITY AND LOGICAL IMPROBABILITY 11

It is no doubt true that Popper recommends such a *prima facie* self-defeating principle. But according to Popper, the recommendation is quite plausible because accepting$_1$ the most improbable hypothesis we are (Popper quotes from William Kneale) accepting$_1$ that one "which we can hope to eliminate most quickly if it is false .... In short, the policy of assuming always the simplest [i.e., for Popper, the most improbable] hypothesis which accords with the known facts is that which will enable us to get rid of false hypotheses most quickly."[29] But the fact that the rule or policy guarantees the *most rapid* growth of knowledge is not its primary strength. Its primary strength is that its denial is self-defeating. In Popper's own words,

There can be no doubt that the absolute probability of a statement *a* is simply the *degree of its logical weakness, or lack of informative content*, and that the relative probability of a statement *a*, given a statement *b* is simply the degree of relative weakness, or the relative *lack* of *new* informative content in statement *a*, assuming that we are already in possession of the information *b*. Thus if we aim, in science, at a high informative content – if the growth of knowledge means that we know more, that we know *a* and *b*, rather than *a* alone, and that the content of our theories thus increases – then we have to admit that we also aim at a low probability... *And since a low probability means a high probability of being falsified*, it follows that a high degree of falsifiability, or refutability, or testability, is one of the aims of science.[30] (Some italics added.)

Third, one might object that by following A, "in the long run" we will have *less* rather than more knowledge about the world. For when we have to choose between two hypotheses, we must, following A, select the one that is most likely to be false given the evidence we now have. So the hypothesis that is most likely to be true is either rejected or just ignored. But if we assume that the more probable hypothesis describes the most likely event and that the more likely event is the one that occurs more often "in the long run," then more often than not our probably-false hypotheses will be falsified and rejected. But this just means that A is such that "in the long run" we shall have fewer and fewer acceptable$_2$ hypotheses. And that means that "in the long run" the rule Popper suggests will lead us to less rather than more knowledge about the world. In reply to this argument, Popper would have to point out that even if it were true that "in the long run" we shall have fewer acceptable$_2$ hypotheses, it does not follow that we shall have less rather than more knowledge about the world. Because

---

[29] William Kneale, *Probability and Induction* (Oxford: Oxford University Press, 1949), p. 229. The quotation is in Popper, *The Logic of Scientific Discovery*, p. 219.
[30] Popper, *Conjectures and Refutations*, p. 219.

our policy, following A, has been to always accept₁ the hypothesis with the greater empirical content, the few hypotheses that survive should be very informative. Furthermore, there is no *logical* reason why we should not have only one acceptable₂ hypothesis (say, perhaps, an extremely powerful theory) from which much more knowledge about the world is derivable than from a number of other hypotheses.

In another attempt to destroy the description argument, one might try to show that the argument is self-contradictory by using the following "proof." Let H be some hypothesis and let e be some evidence gathered in support of H. It is assumed that e itself is acceptable and logically equivalent to H. Hence,

$$p(H, e) = p(e, e) = 1$$

But that means, given e, H tells us nothing about the world at all. So, H is unacceptable. But H is logically equivalent to e. So, e is unacceptable. Thus, e is both acceptable and unacceptable, which is impossible.

The plausibility of this argument (if it has any at all) is completely shattered once subscripts are added to 'acceptable' and 'unacceptable.' Then it runs thus: Let H be some hypothesis and let e be some evidence gathered in support of H. We assume that e itself is acceptable₂. Now suppose e is logically equivalent to H. In that case we have

$$p(H, e) = p(e, e) = 1$$

But that means, given e, H tells us nothing about the world at all. So, H is unacceptable₁. But H is logically equivalent to e. So, e is unacceptable₁. Thus, e is both acceptable₂ and unacceptable₁, which is clearly *not* impossible. It is, after all, the *initial* acceptability₂ of e that is high and the *relative* acceptability₁ of e that is low. The *initial* acceptability₂ of e is just the acceptability₂ of e. The *relative* acceptability₁ of e is the acceptability₁ of e *given* e. It is quite common to have the initial and relative probabilities of a sentence differ, as it is common to have the probabilities of a hypothesis differ relative to different bodies of evidence.[31] Thus, if e is initially acceptable₂ then

---

[31] A thorough examination of such cases may be found in C. G. Hempel, "Inductive Inconsistencies," *Logic and Language: Studies Dedicated to Professor Rudolf Carnap on the Occasion of His 70th Birthday* (Dordrecht, Holland: D. Reidel Publishing Co., 1962), pp. 128–139.

relative to itself, e can tell us nothing about the world; that is why e is relatively unacceptable$_1$.

Finally, consider a case in which the hypotheses H and K are supposed to be used to explain or account for some phenomena described by M. The empirical content of K is less than that of H but the *relevance* of K to M is greater than the relevance of H to M. In fact, in the case we are considering, H is totally irrelevant to M. Now, in such a situation it is apparent that K would be a more acceptable$_1$ explanatory hypothesis than H (i.e., because H is irrelevant to M). Thus we have a case in which the more acceptable$_1$ hypothesis is *not* the one with the greater empirical content. The following example proves that such situations are possible. First, we define a relevance function r.

$$r(M, H) \text{ if, and only if } p(M, H) \gtreqless p(M)$$

That is, a hypothesis H is relevant to a sentence M if, and only if, the probability of M given H *differs* from the probability of M.[32] Let K be 'x is a cat,' H is 'x is a dog,' L is 'x is a bird,' M is 'x is domesticated' and N is 'x is wild.' We will need one domesticated and three wild dogs, two domesticated and eight wild cats, and one domesticated and one wild bird. In tabular form we have the following distribution.

		domesticated M	wild N
H	dog	1	3
K	cat	2	8
L	bird	1	1

Now, we have

$$p(M, H) = \tfrac{1}{4} = p(M)$$

So, H is irrelevant to M. And we have

$$p(M, K) = \tfrac{1}{5} < p(M) = \tfrac{1}{4}$$

So, K is relevant to M. But

$$p(H) = \tfrac{1}{4} < p(K) = \tfrac{5}{8}$$

which means

$$Ct(H) > Ct(K).$$

[32] This is Carnap's definition of relevance. See Carnap, *The Logical Foundations of Probability*, p. 348; below Chapter III, Section 4 and Chapter IV, Sections 2 and 3.

Therefore, we have a case in which although the empirical content of H is greater than that of K, the relevance of K to the phenomena to be explained M is greater than the relevance of H to M. And in this case the hypothesis with the smaller empirical content would be more acceptable$_1$.

Now it seems to me that Popper might want to meet this challenge with the following reply. We must grant that one cannot successfully defend the view that a high degree of empirical content (low logical probability) is a *sufficient* condition of acceptability$_1$. The case we have just considered leaves no doubt about that. However, one might still claim that *if all other things are equal* (e.g., relevancy, explanatory power, simplicity, risk of errors, etc.) then the hypothesis with the lowest probability will be the most acceptable$_1$. For example, with respect to the case above, one might say that if the relevance of H and K to M had been equal then H would have been more acceptable$_1$ because its empirical content was greater than that of K.

When I wrote my doctoral dissertation in 1964, I thought that this was a satisfactory way to save the main thrust of Popper's argument. But since that time, as I have suggested elsewhere,[33] I have come to doubt the wisdom of such a move. The fact is that the *ceteris paribus* condition logically cannot be met because the risk of error (probability) is, as Popper insists, complementary to the empirical content of a hypothesis. One is always confronted by a choice between staying close to one's data on the one hand and learning more about the world on the other; between, if you like, the safety of the known and the excitement of the unknown. Notice, however, that this tension is completely missing in the description argument, and this is its fatal defect. What one ought to say is something like the following sixth and I think decisive objection.

Because *one* of the goals of a scientist is the discovery of laws and theories that adequately described the world in which we live, he ought to search for "bold" hypotheses. But because *another* of the goals of a scientist is *efficient* or *prudential* inquiry, he ought to make the most efficient use of all of the resources at his command, including all of the knowledge that he thinks he has. It is this aspect of scientific inquiry that the description argument entirely ignores. Hence, the

---

[33] Alex C. Michalos, "Positivism versus the hermeneutic-dialectic school," *Theoria*, XXXV (1969), pp. 267–278.

argument involves an informal fallacy generally known as 'special pleading.' It leads us to a false conclusion through the omission of important information about other goals of scientists. Given the dual goal of *efficient growth* or *prudential growth*, it is apparent that the most acceptable$_1$ hypothesis is the one which provides the greatest amount of "boldness" that is consistent with the most tolerable or wisest margin of safety when all aspects of the research programme have been carefully examined. To quote from an earlier paper: "As long as there is *some* chance of making a mistake, the necessary condition of growth is satisfied. As long as the odds in favor of success are good, a necessary condition of *prudential* growth is satisfied."[34]

---

[34] *Ibid.*, p. 277.

CHAPTER III

TWO EXPLICANDA AND THREE ARGUMENTS

1. INTRODUCTION

In this chapter we shall consider three arguments that share one common and erroneous assumption.[1] The arguments appeared separately in 1954,[2] 1956[3] and 1959,[4] and will be considered in that order. In section two I distinguish briefly three kinds of concepts, namely, classificatory, comparative and quantitative concepts. These distinctions are necessary for a clear understanding of the arguments in the other sections. In section three I present Popper's argument that probability must not be identified with confirmation because a measure of confirmation cannot satisfy certain rules of the Probability Calculus.[5] Section four contains a *reductio ad absurdum* refutation of Popper's 1954 argument. Sections five and seven contain Popper's 1956 and 1959 arguments respectively. *Reductio ad absurdum* refutations of each of these arguments are constructed in sections six and eight. Although the *reductio ad absurdum* arguments must be conclusive as far as Popper is concerned (because they are based on Popper's own assumptions) they are not otherwise admissible. They are not otherwise admissible because one of Popper's assumptions is plainly false (as shown in section ten). However, in sections nine through eleven the view of J. G. Kemeny[6] and Carnap[7] regarding the three

---

[1] See sections 9–11 below.
[2] Popper, "Degree of Confirmation," *loc. cit.*, pp. 143–149. The note is reprinted in Appendix *ix of Popper, *The Logic of Scientific Discovery*, pp. 395–402 and our page references are to this latter version.
[3] Popper, "Adequacy and Consistency: A Second Reply to Dr. Bar-Hillel," *loc cit.*, pp. 249–256.
[4] Popper, *The Logic of Scientific Discovery*, pp. 387–394.
[5] Popper, "Degree of Confirmation," *loc. cit.*, pp. 395–399 in *The Logic of Scientific Discovery*.
[6] J. G. Kemeny, "Review of K. R. Popper 'Degree of Confirmation,'" *Journal of Symbolic Logic*, XX (1955), 304–305.
[7] Rudolf Carnap, "Remarks on Probability," *Philosophical Studies*, XIV (1963), 65–75; and "Replies and Systematic Expositions," *loc. cit.*, pp. 995–998.

arguments is elaborated and defended. The arguments considered in these sections are surely conclusive. In particular it is shown that Popper and Carnap are concerned with two different *explicanda*. Once this is clearly demonstrated, the *second* main source of dispute between Popper and Carnap is exposed and easily eliminated. It is eliminated in these sections by showing that because two *explicanda* are involved, Popper's arguments must lead to trivial conclusions. Indeed, his alternatives are clearly to construct arguments with quite trivial conclusions *or else* to construct no arguments (i.e., relevant to the issues here) at all.

2. THREE KINDS OF CONCEPTS[8]

It is customary and it will be useful to distinguish three kinds of concepts. The concepts represent three ways to group or order things (e.g., facts, objects, people, etc.). Suppose we are given the task of separating a class of children into any number of mutually exclusive and exhaustive groups. We are permitted to use any criterion we choose to construct the groups. For example, those children with brown hair are put into Group I and those without brown hair are put into Group II. If D represents the total number of children in the class, we have $I + II = D$, $D - I = II$ and $D - II = I$. That is, we have classified D into two mutually exclusive and exhaustive classes, namely, Group I and Group II. Group I and Group II are *classificatory* concepts, i.e., concepts used to group objects into separate (and in this example, mutually exclusive and exhaustive) classes. We could have, of course, considered red, blond and black hair (instead of just nonbrown hair) and constructed four smaller classes in the same fashion.

Alternatively, we might draw one horizontal line on the wall, march the children by it and put all those whose ears are higher than the line in one class and all those whose ears are lower than the line in another class. The two groups thus formed are separated on the basis of the relation of their ears to the line. If x's ears are higher than the line, he goes into Group A; otherwise he goes into Group B. Group A and Group B are formed by applying the comparative concepts

---

[8] A more thorough discussion of these concepts may be found in Carnap, *The Logical Foundations of Probability*, pp. 8–11. and in Carl. G. Hempel, "Fundamentals of Concept Formation in Empirical Science," *International Encyclopedia of Unified Science*, II (1960), 50–79.

'higher than' and 'lower than.' *Comparative* concepts indicate not only grouping (as classificatory concepts) but order too. That is, they are used to express the fact that certain objects possess more or less of a particular property. In the case above, for instance, the property in question is height.

Finally, we might separate the class of children into smaller groups or subclasses by applying a concept used to indicate not only order but distance or number of units from some point. That is, we might construct subclasses by applying a metrical or *quantitative* concept. We might, for example, correlate a certain length of a straight rod with the word 'inch' and simply count the number of inches that must be put side by side to reach a length equal to the length (i.e., height) of each child. Using the quantitative concept then (i.e., the concept of an inch) we can determine *how much* higher than the line some children are and how much lower than the line others are. We can determine how much taller or shorter each child is than all the others. Given a quantitative concept, a number of such interesting questions become answerable.

For our purposes, it is important to point out that it is possible to distinguish three kinds of concepts of probability or confirmation. One might, that is, explicate a classificatory, comparative, or quantitative concept of, say, confirmation. It is important to make these distinctions apparent because some questions are applicable when one sort of concept is used but not when another is used. For example, if someone tells us that event x is probable in a classificatory sense then it may be pointless to ask, How probable?[9] And if the same person tells us in the same sense that y is also probable, it may be pointless to ask, How probable is the disjunction of x and y? On the other hand, if x and y were probable in the quantitative sense then by the Addition Rule of the Probability Calculus a perfectly straightforward reply to 'How probable is the disjunction of x and y?' might be offered.

---

[9] That is, it might be the case that either there is no known way to measure the degree of probability of x or there is no known way to even compare the probability of x with something else. In Chapter III of *A Treatise on Probability* (London: Macmillan and Co., 1921), John Maynard Keynes considers both kinds of cases. His view is: "Whether or not such a thing is theoretically conceivable, no exercise of the practical judgment is possible, by which a numerical value can actually be given to the probability of every argument. So far from our being able to measure them, it is not even clear that we are always able to place them in an order of magnitude. Nor has any theoretical rule for their evaluation ever been suggested" (pp. 27–28).

3. POPPER'S 1954 ARGUMENT

By the term 'Calculus of Probability' we mean an axiomatic system with the following equations as axioms or provable theorems. If x and y represent any hypothesis whatever, then

1. *Non-negative Value Rule*: $p(x) \geq 0$.
2. *Summation Rule*: $\sum_{i=1}^{n} p(x_i) = 1$ for all of the n mutually exclusive and exhausive hypotheses in a given universe of discourse.
3. *Special Addition Rule*: $p(x \vee y) = p(x) + p(y)$ provided that $x \cdot y$ is logically false.
4. *General Multiplication Rule*: $p(x \cdot y) = p(x)p(y, x) = p(y)p(x, y)$.[10]

In 1954 Popper tried to show that any function that satisfied the third and the fourth of these equations could not be an adequate corroboration or confirmation function. Here is an outline of his argument.

Because probability values are, by convention, such that

$$0 \leq p(x, z) \leq 1$$

(i.e., the probability of x on z is 0, 1 or a fraction), according to the General Multiplication Rule

(1) $$p(x, z) \geq p(xy, z).$$

According to the Special Addition Rule and an elementary distribution rule of logic

(2) $$p(xy \vee x\bar{y}, z) \geq p(xy, z)$$

because

$$(xy \vee x\bar{y}) \equiv [x \vee (y\bar{y})] \equiv x.$$

Thus both rules have (1) as a consequence and that means that (1) may be taken as a general characteristic of the probability functor. Now Popper tries to prove that an adequate corroboration functor C may violate (1). That is, he tries to prove that there are cases in which

(3) $$C(xy, z) > C(x, z)[11].$$

---

[10] For a more thorough presentation of the elements of the calculus of probability see my *Principles of Logic*, Chapter Six.

[11] "C(x,z)" is read "the degree of confirmation (or corroboration) of x by z". Throughout this essay I make trivial substitutions without noting them, to simplify the notation. For example, here I substitute "z" for "y" in Popper's notation so I can use the simpler "xy" instead of x_1x_2 later. I also write "p" instead of "P" throughout.

If that is true then clearly

(4) $\qquad C(x, z) \neq p(x, z)$, or simply $C \neq p.$[12]

To prove (4) he begins by defining a classificatory concept of support or confirmation. Letting Co(x, z) stand for 'z confirms (or supports) x' and not Co(x, z) stand for 'z disconfirms (or undermines) *or* is independent of (or irrelevant to) x' he suggests the following definitions.

(5) $\qquad Co(x, z) \underset{df}{=} p(x, z) > p(x)$

(6) $\qquad \text{not } Co(x, z) \underset{df}{=} p(x, z) \leq p(x)$

He claims, that is, that "the power of z to support [or confirm] x ... is essentially a *measure of the increase or decrease* due to z, in the probability of x."[13] Then he claims that if z supports x but undermines y then "we should have to say in such a case that z confirms x to a higher degree than it confirms y."[14] That is,

(7) $\qquad$ if $Co(x, z)$ and not $Co(y, z)$, then $C(x, z) > C(y, z)$.

With these stipulations and the following example, we may prove (4). We have four colored counters a, b, c, and d with four exclusive and equally probable properties, blue, green, red and yellow. Let x be 'a is blue or green,' y is 'a is blue or red' and z is 'a is blue or yellow.' We have

(8) $\qquad p(z) = p(y) = p(x) = \frac{1}{2} = p(x, z)$

and

(9) $\qquad p(xy) = \frac{1}{4} < p(xy, z) = \frac{1}{2}.$

According to (5) and (9), and (6) and (8)

(10) $\qquad Co(xy, z)$ and not $Co(x, z)$.

That gives us (3) according to (7). Thus, because we have (3) but could never have

(11) $\qquad p(xy, z) > p(x, z)$

according to (1) and (2), it follows that $C \neq p$. That is, (4) must be true.[15]

---

[12] Popper, *The Logic of Scientific Discovery*, pp. 396–397.
[13] *Ibid.*, p. 399.
[14] *Ibid.*
[15] *Ibid.*, p. 398.

We may then, summarize Popper's argument thus. If we identify probability and corroboration (or confirmation) then the General Multiplication and Special Addition rules of the Probability Calculus must be applicable to the latter. But the truth of (3) proves that the probability rules are not applicable to confirmation. Thus, we must not identify probability and confirmation.[16]

4. REDUCTIO AD ABSURDUM OF THE 1954 ARGUMENT

It is apparent that the truth of (7) is required for the 1954 argument. Popper claims that (7) is tautologous. In 1956 he wrote

> My assertion that [7] is tautologous, or analytic, is not merely based upon intuitive grounds. It is based upon the fact that *every statement* is true which results from substituting in [7] a grammatically fitting predicate for 'confirmed'... for [7] is an instance of the general schema: If x does not possess the property P, and y does possess the property P, then it cannot happen that x possesses the property P in the same degree as y, or in a higher degree than y ... *This schema applies ... to ...* 'y *is probable, given z.*'[17] (Italics added)

To apply the schema to "y is probable, given z" we would begin by defining a classificatory concept of probability. Let pr(x, z) stand for 'x is probable, given z.' Then following (5) and (6)

(5p) $$\text{pr}(x, z) \underset{df}{=} p(x, z) > p(x)$$

and

(6p) $$\text{not pr}(x, z) \underset{df}{=} p(x, z) \leq p(x).$$

Obviously pr(x, z) is equivalent to Co(x, z) since they are defined exactly alike. (Incidentally, both are equivalent to Carnap's concept of *positive relevance*.)[18] There is nothing extraordinary here. If it is legitimate to define a classificatory concept of *confirmation* in terms of a quantitative concept of probability, it must *a fortiori* be legitimate

---

[16] It is worthwhile to mention that at least four other writers have found the General Multiplication Rule inapplicable for a Calculus of Confirmation. See C. G. Hempel and P. Oppenheim, "A Definition of 'Degree of Confirmation'," *Philosophy of Science*, XII (1945), 98–115 (especially section 10); Olaf Helmer and Paul Oppenheim, "A Syntactical Definition of Probability and of Degree of Confirmation," *Journal of Symbolic Logic*, X (1955), 25–60 (especially sections 9 and 11); R. H. Vincent, "A Note on Some Quantitative Theories of Confirmation," *Philosophical Studies*, XII (1961), 91–92 and "On My Cognitive Sensibility," *ibid.*, XIV (1963), 77–79.

[17] Popper, "Adequacy and Consistency: A Second Reply to Dr. Bar-Hillel," *loc. cit.*, pp. 254–255.

[18] See Carnap, *The Logical Foundations of Probability*, p. 348.

to define a classificatory concept of probability in terms of a quantitative concept of probability.

But with (5p), (6p) and

(7p) if pr(x, z) and not pr(y, z), then p(x, z) > p(y, z)

we may reduce the 1954 argument to absurdity. By (5p) and (9), and (6p) and (8) we have

(10p) pr(xy, z) and not pr(x, z).

That gives us (the impossible !)

(11) p(xy, z) > p(x, z)

according to (7p). Thus, because we must both assert and deny (11) if we grant that p = p, we must deny that p = p!

We may then, summarize my formally identical version of Popper's argument thus. If we identify the quantitative concept of probability with itself (i.e., assert the trivially true p = p) then the General Multiplication and Special Addition rules of the Probability Calculus must be applicable to this concept. But the "truth" of (11) (11 was "proven" using Popper's argument but substituting 'probability' for 'confirmation' throughout) "proves" that the probability rules are not applicable to probability! Thus, we must not identify probability with probability, i.e., we must deny p = p.

The *reductio ad absurdum* refutation of Popper's argument then, runs as follows. If we accept his argument as proof that "degree of confirmation" does not equal "degree of probability" (i.e., $C \neq p$) then we must accept my argument as proof that "degree of probability" does not equal itself (i.e., $p \neq p$). But p = p. Thus, my argument must be rejected, and therefore so must Popper's.

Let me again reconstruct the main lines of my argument. Popper's argument rests on the truth of (7). To support his claim that (7) must be true, he makes the stronger claim that (7) is simply an instance of a tautologous schema. Thus, if we do not grant the truth of this stronger claim, we have no support for this claim that (7) must be true. So, we grant the truth of the stronger claim. But then we must grant the truth of (7p). And this allows us to construct the argument leading to the absurd consequence that $p \neq p$. Finally, since this absurdity is obtained by using an argument that is formally identical to Popper's argument leading to $C \neq p$, we must reject *both* arguments.

5. THE 1956 ARGUMENT

In his 1956 article[19] Popper argues that if one identifies the quantitative concept of probability with degree of confirmation then one may derive the following invalid theorem.

If x follows from y, then, for every z,
if $p(y, z) > p(y)$ then $p(x, z) > p(x)$.[20]

That is,

(16) $\quad (y \supset x) \supset (z)[(p(y, z) > p(y)) \supset (p(x, z) > p(x))]$.

We may prove (16) is invalid thus. Consider the next throw of a homogeneous die. Let x be 'an even number turns up,' y is 'the face with only two dots turns up' and z is 'a face with a number less than five turns up.' x follows from y because if number two turns up, an even number turns up. But we also have

$$p(y, z) = \tfrac{1}{4} > p(y) = \tfrac{1}{6}$$

and

$$p(x, z) = \tfrac{1}{2} = p(x)$$

which is incompatible with (16).

Popper claims then, that given

(7) $\quad (Co(y, z) \cdot {-}Co(x, z)) \supset (C(y, z) > C(x, z))$[21]

and

(17) $\quad (y \supset x) \supset (z) - (C(y, z) > C(x, z))$[22]

we may derive (16); and since (7) is supposed to be tautologous, (17) must be rejected even though its analogue for the probability calculus

$$(y \supset x) \supset (z) - (p(y, z) > p(x, z))$$

is valid. Here is how we would derive (16) from (7) and (17).

(18) $\quad -(C(y, z) > C(x, z)) \supset -(Co(y, z) \cdot {-}Co(x, z)) \quad$ [Trans. (7)]

---

[19] Popper, "Adequacy and Consistency: A Second Reply to Dr. Bar-Hillel," *loc. cit.*, pp. 254–256.
[20] *Ibid.*, p. 254.
[21] Notice this formula is only (7) with x and y interchanged.
[22] (17) is equivalent to T59–2d, p. 317, in Carnap, *The Logical Foundations of Probability*. I have only simplified the formula and the following derivation by writing it in its present form. The more important point is that Carnap accepts (17) and Popper does not.

(19) $(y \supset x) \supset -(Co(y,z) \cdot -Co(x,z))$ [(17) U.I., (17) and (18) H.S.]

(20) $(y \supset x) \supset (z)(Co(y,z) \supset Co(x,z))$ [(19) Imp., U.G.]

(21) $(y \supset x) \supset (z)[(p(y,z) > p(y)) \supset (p(x,z) > p(x)]$
[Subst. by (5)]

6. REDUCTIO AD ABSURDUM OF THE 1956 ARGUMENT

Here, as in the case of the 1954 argument, it is possible to construct an argument that is formally identical to the 1956 argument but has the absurd consequence that $p \neq p$. We take as our assumptions the new instance of Popper's alleged tautologous schema

(7p) $(pr(y,z) \cdot -pr(x,z)) \supset (p(y,z) > p(x,z))$

and

(17p) $(y \supset x) \supset (z) - (p(y,z) > p(x,z))$.

And the derivation of (16) runs thus.

(18p) $-(p(y,z) > p(x,z)) \supset -(pr(y,z) \cdot -pr(x,z))$
[Transposing (7p)]

(19p) $(y \supset x) \supset -(pr(y,z) \cdot -pr(x,z))$
[(17p) U.I., (17p) and (18p) H.S.]

(20p) $(y \supset x) \supset (z)(pr(y,z) \supset pr(x,z))$
[(19p) Imp., U.G.]

(21p) = (16) $(y \supset x) \supset (z)[(p(y,z) > p(y)) \supset (p(x,z) > p(x))]$
[Substitution by (5p)]

Thus, if we identify p with itself (i.e., grant that $p = p$) we are able to derive the invalid (16). So, we must deny that $p = p$.

The *reductio ad absurdum* refutation of Popper's 1956 argument then, runs as follows. If we accept his argument as proof that the identification of "degree of confirmation" with "degree of probability" (i.e., $C = p$) allows us to derive the invalid theorem (16), and that therefore we must deny that $C = p$, then we must accept my argument as proof that the identification of the quantitative concept of probability with itself (i.e., $p = p$) allows us to derive the invalid theorem (16), and therefore we must deny $p = p$. But $p = p$. Thus, my argument must be rejected, and therefore so must Popper's.

7. AN ALLEGED CONTRADICTION

In 1959 Popper claimed that "Carnap's theory is self-contradictory, and that its contradictoriness is not a minor matter which can be easily repaired, but is due to mistakes in its logical foundations."[23]

Popper notes again, that Carnap accepts the definitions of the classificatory concepts of confirmation and disconfirmation offered above. Then he quotes the following remark from Carnap about a sentence that is formally identical to the negation of (7), i.e. briefly (-7). "If the property Warm and the relation Warmer were designated by ..., say, 'P' and 'R,' then 'Pa · −Pb · Rba' would be self-contradictory."[24] Finally, Popper attempts to prove that the above sentence (i.e., (-7)) may be derived from Carnap's system by the following case.[25] Consider the next throw of a homogeneous die. Let x be the statement 'six will face up'; y is its negation and z is 'an even number will face up.' We have then:

$$p(x) = \tfrac{1}{6}, p(y) = \tfrac{5}{6}, p(z) = \tfrac{1}{2}$$

(22) $$p(x, z) = \tfrac{1}{3} > p(x)$$

and

(23) $$p(y, z) = \tfrac{2}{3} > p(y).$$

But

(24) $$p(x, z) > p(y, z).$$

Hence, according to our definitions (5) and (6), if we substitute C for p in (24), we find that the conjunction of (22), (23) and (24) is just the self-contradictory (-7).

8. REDUCTIO AD ABSURDUM OF THE ARGUMENT FOR THE ALLEGED CONTRADICTION

If we assume again

(5p) $$\text{pr}(x, z) \underset{\text{df}}{=} p(x, z) > p(x)$$

and

(6p) $$\text{not } \text{pr}(x, z) \underset{\text{df}}{=} p(x, z) \leq p(x)$$

---

[23] Popper, *The Logic of Scientific Discovery*, p. 393.
[24] *Ibid.*
[25] *Ibid.*, pp. 390–391.

and that

(−7p) $\text{pr}(x, z)$ and not $\text{pr}(y, z)$ and $p(x, z) < p(y, z)$

is self-contradictory (again granting Popper's claim that (7p) is tautologous and that therefore its denial (−7p) is self-contradictory), we may construct an argument formally identical to the 1959 argument. According to (22) and (5p), and (23) and (6p) we have

$\text{pr}(x, z)$ and not $\text{pr}(y, z)$.

Conjoining this to (24), we have the self-contradictory (−7p).

The situation then, is as follows. Popper argues that if we assume $C = p$, (5) and (6), we must assert the self-contradictory (−7). Thus, we should deny that $C = p$. By a formally identical argument I show that the assumption of (5p), (6p) and $p = p$, forces us to assert (−7p). Thus, we should deny that $p = p$.

The *reductio ad absurdum* refutation of Popper's argument then, runs as follows. If we accept his argument as proof that the identification of "degree of confirmation" with "degree of probability" (i.e., $C = p$) leads us to the self-contradictory (−7), and that therefore we must deny $C = p$, then we must accept my argument as proof that the identification of "degree of probability" with itself (i.e., $p=p$) leads us to the self-contradictory (−7p), and therefore we must deny that $p = p$. But $p = p$. Thus, my argument must be rejected, and therfore so must Popper's.[26]

9. JOHN G. KEMENY'S REVIEW OF THE 1954 ARTICLE

In a review of Popper's article written for the *Journal of Symbolic Logic*,[27] J. G. Kemeny wrote the following.

A careful reading of this paper will show that *Popper and Carnap have two different explicanda* in mind. Carnap's can be described as 'how sure we are of x if we are given z as evidence,' while Popper's concept is 'how much surer we are of x given z than without z.' Hence most of the present paper is devoted to a de-

---

[26] It is perhaps, worth mentioning that R. H. Vincent has replied to the 1959 argument in roughly the same fashion in "Concerning An Alleged Contradiction," *Philosophy of Science*, XXX (1963), 189–194. Carnap called my attention to this article in a letter written to me (May, 1964) in response to a note (roughly identical to Section 8 below) I sent him (Carnap) regarding the 1959 argument. Carnap was satisfied with my argument and wrote "I think your argument is correct and convincing. R. H. Vincent has published a note on the same point ... but it seems to me that your argument is stronger."

[27] Kemeny, *loc. cit.*, pp. 304–305.

monstration that Carnap's explicatum of the former concept is not an adequate explicatum of the latter concept, a fact that Carnap would be the first to admit.[28]

In reply to this, Popper wrote

In fact 'a careful reading' of my paper – and, I should add, of Carnap's book – will *not* 'show that Popper and Carnap have two different *explicanda* in mind,' but it will show that Carnap had inadvertently two different and incompatible 'explicanda' in mind with his probability$_1$, one of them my C, the other my p.[29]

In this section I will show that Popper's reply is mistaken. It is significant to note that Carnap and Y. Bar-Hillel have roughly the same view as Kemeny regarding the two explicanda.[30]

To begin with then, we will follow Carnap's precise discussion of the two explicanda to which Kemeny refers. Carnap claims that the prescientific or presystematic expression 'z confirms x' is ambiguous.[31] In particular it might mean either of the following different things: 'x is made "firm" or probable by z' *or* 'x is made "firmer" or *more* probable by z.' It is apparent that x is made *more* probable by z if, and only if, the probability of x given z is greater than the probability of x without z. Thus, it is equally apparent that when Popper defined 'z confirms x' in (5), he was defining 'x is made "firmer" or *more* probable by z.' That this was indeed Popper's intention is further supported by his remark that "the power of z to support x ... is essentially a *measure of the increase or decrease* due to z, in the probability of x."[32] The classificatory, comparative and quantitative concepts corresponding to this *explicandum* may be expressed as follows.

Popper's Three Concepts[33]

1. Classificatory concept: The probability of x given z is
   z confirms x.   greater than the probability of x
                             alone; i.e., $p(x, z) > p(x)$.

---

[28] *Ibid.*, p. 304.
[29] Popper, *The Logic of Scientific Discovery*, p. 393.
[30] Carnap, "Replies and Systematic Expositions," *loc. cit.*, and Carnap, "Remarks on Probability," *loc. cit.*, Y. Bar-Hillel, "Comments on 'Degree of Confirmation' by Professor K. R. Popper," *B.J.P.S.*, VI (1955), 155–157 and Y. Bar-Hillel, "Further Comments on Probability and Confirmation, A Rejoinder to Professor Popper," *B.J.P.S.*, VII (1956), 245–248.
[31] Carnap, "Remarks on Probability," *loc. cit.*, pp. 68–72.
[32] Popper, *The Logic of Scientific Discovery*, p. 399.
[33] Lest the reader be misled by this title, I should point out that I am using this title as an abbreviation for "Definitions of the three concepts of confirmation using 'z confirms x' in the sense of 'z makes x firmer or *more* probable.'"

2. Comparative concept: The probability of x given z is
   z confirms x more than z' con- greater than the probability of x
   firms x'. alone *more than* the probability of
   x' given z' is greater than the
   probability of x' alone; i.e.,
   $p(x, z) > p(x)$ *and* $p(x', z') > p(x')$
   and $p(x, z) - p(x) > p(x', z') - p(x')$

3. Quantitative concept: The probability of x given z is
   The degree of confirmation of greater than the probability of x
   x given z is u. alone to a degree u; i.e.,
   $p(x, z) > p(x)$ *and*
   $p(x, z) - p(x) = u$.

On the other hand, the classificatory, comparative and quantitative concepts corresponding to Carnap's *explicandum* (*if* he had used only one) may be expressed thus.

Carnap's Three Concepts[34]

1. Classificatory concept: The probability of x given z is
   z confirms x. greater than some fixed number
   $u \geq 0$; i.e., $p(x, z) > u \geq 0$.
2. Comparative concept: The probability of x given z is
   z confirms x more than z' con- greater than the probability of
   firms x'. x' given z'; i.e., $p(x, z) > p(x', z')$.
3. Quantitative concept: The probability of x given z is
   z confirms x to a degree u. u; i.e., $p(x, z) = u$.

Carnap has admitted that in *The Logical Foundations of Probability* he explicated Popper's *classificatory* concept of confirmation instead of his own.[35] That is, instead of explicating what I have referred to as Carnap's Three Concepts, Carnap explicated Popper's first concept and Carnap's second in terms of Carnap's third; i.e., a classificatory and comparative concept of confirmation were explicated in terms of a quantitative concept.

Now Popper's paper and Carnap's book are primarily concerned with the *quantitative* concept of confirmation. Thus, Kemeny's remarks

---

[34] This is also an abbreviation for the other explicandum. See the last note.
[35] Carnap, "Remarks on Probability," *loc. cit.*, pp. 68–72.

must be taken as claims about the quantitative concept 'degree of confirmation,' and with respect to *that* concept, it is apparent that Popper and Carnap *have* "different explicanda in mind." Carnap wrote, "the chief topic of this book is the problem of an explication of probability" (i.e., the logical concept of probability).[36] Carnap's name for the explicatum of the quantitative concept of logical probability (the explicandum) is 'degree of confirmation.'[37] Furthermore, among Carnap's conditions of adequacy for his explicatum we may *always* find the Special Addition and General Multiplication rules;[38] and that means he always intended his 'degree of confirmation' to be an explicatum for *some* concept of probability. Therefore, it is false that by 'probability' (in its quantitative form) Carnap meant Popper's 'degree of confirmation' part of the time and Popper's 'logical probability' the rest of the time.

10. FINAL APPRAISAL AND REFUTATION OF THE 1954 ARGUMENT

We must now consider the last sentence in the Kemeny quotation with regard to its adequacy as an appraisal of Popper's 1954 argument. The question is: Has Popper *only* demonstrated that "Carnap's explicatum of [degree of probability] is not an adequate explicatum of [degree of difference in probability] ...?"[39] I shall defend an affirmative answer to this question. More precisely, in this section, I shall show exactly where Popper goes astray in his 1954 argument and in the following section I shall show that the same confusion occurs in his 1956 and 1959 arguments.

To begin with, you will recall that Popper's arguments rest on three assumptions.

(5) $$\text{Co}(x, z) \underset{df}{=} p(x, z) > p(x)$$

(6) $$\text{not } \text{Co}(x, z) \underset{df}{=} p(x, z) \leq p(x)$$

and

(7) If $\text{Co}(x, z)$ and not $\text{Co}(y, z)$, then $C(x, z) > C(y, z)$

---

[36] Carnap, *The Logical Foundations of Probability*, p. 163.
[37] *Ibid.*
[38] See for example, Carnap, *The Logical Foundations of Probability*, p. 285; R. Carnap, *The Continuum of Inductive Methods* (Chicago: University of Chicago Press, 1952), p. 12; Carnap, "Replies and Systematic Expositions," *loc. cit.*, p. 974.
[39] It might be useful for the reader to refer to Kemeny's remarks again at the beginning of Section 9.

Carnap accepts (5) and (6) as definitions of workable *classificatory* concepts of confirmation, and disconfirmation or independence. He claimed "the classificatory concept of confirmation as used, for instance, by a scientist when he says something like this: 'The result of the experiment just made supplies confirming evidence for my hypothesis,'" should be defined (in terms of the quantitative concept of probability) according to (5).[40] Thus, the problematic premise in Popper's argument must be (7). (7) is crucial for Popper's derivation of an invalid law (1956)[41] and for his 1959 charge of inconsistency.[42] In the 1956 derivation (Sec. 5) (5) and (7) are assumed true and in the 1959 charge of inconsistency (Sec. 7), the denial of (7) is assumed self-contradictory.

We should point out first, that (7) is not *formally* a logical truth. Formally it is only an instance of

If x has the relation R to z *and* y does not have the relation R to z *then* x has the relation S to z more than y has the relation S to z.

That is, simply

if Rxz *and* $-$Ryz, *then* Sxz $>$ Syz

which is not a logical truth. However, like other contingent schemata, it does admit of *interpretations* that are always and/or necessarily true. Herein lies Popper's confusion.

Let us agree to let 'Cc' stand for Carnap's concept of 'degree of confirmation' and 'Cp' for Popper's. Thus, the explicandum of 'Cp' is roughly *'difference* in probability' and the explicandum of 'Cc' is simply *'probability.'* The denial of (7) is:

($-$7)   Co(x, z) and not Co(y, z) and C(x, z) $<$ C(y, z).

Now Carnap claims "it [$-$7] is indeed absurd if we *interpret* C in the sense of Cp. Since my [Carnap's] concept Cc is the same as logical probability, the statement [$-$7] *interpreted* in the sense of Cc is not at all absurd."[43] (Italics added.) We may, in fact, interpret [$-$7] in at least three ways and of these three, only the first two are absurd. Carnap claims, and it is indeed *true*, that the third interpretation is

---

[40] Carnap, *The Logical Foundations of Probability*, p. 475.
[41] Popper, "Adequacy and Consistency: A Second Reply to Dr. Bar-Hillel," *loc. cit.*, pp. 254–256.
[42] Popper, *The Logic of Scientific Discovery*, pp. 387–394.
[43] Carnap, "Replies and Systematic Expositions," *loc. cit.*, p. 998.

not absurd. Using our new notation, the three interpretations are:

(−7a) Co(x, z) and not Co(y, z) and Cp(x, z) < Cp(y, z)

(−7b) *Co*(x, z) and not *Co*(y, z) and Cc(x, z) < Cc(y, z)

(−7c) Co(x, z) and not Co(y, z) and Cc(x, z) < Cc(y, z)

In (−7b) '*Co*(x, z)' represents the classificatory concept of confirmation according to our description in the table of Carnap's Three Concepts. The explicandum in (−7a) is '*difference* in probability'; the explicandum in (−7b) is 'probability'; and *both* of these explicanda occur in (−7c). To make these more explicit I shall express them using only the functor p, i.e., I write them in terms of the quantitative concept of logical probability.

(−7a) $p(x, z) > p(x)$ and $p(y, z) \leq p(y)$ and
         $p(x, z) - p(x) < p(y, z) - p(y)$

(−7b) $p(x, z) > 0$ and $p(y, z) = 0$ and $p(x, z) < p(y, z)$

(−7c) $p(x, z) > p(x)$ and $p(y, z) \leq p(y)$ and $p(x, z) < p(y, z)$

Now, (−7a) must be impossible because $p(x, z) - p(x)$ will be *positive* when $p(x, z) > p(x)$, and $p(y, z) - p(y)$ will be *negative or o* when $p(y, z) \leq p(y)$. And, (−7b) is obviously impossible. But (−7c) is quite possible as the following case proves. We have ten children; one boy and three girls under two years of age, and one boy and five girls over two years of age. Let x be 'person 0 is a boy,' y is '0 is a girl' and z is '0 is under two years of age.' In tabular form we have:

		x	y
		boys	girls
z	under two	1	3
$\bar{z}$	over two	1	5

All our conditions are met since

$$p(x) = \tfrac{1}{5} < p(x, z) = \tfrac{1}{4}$$
$$p(y) = \tfrac{4}{5} > p(y, z) = \tfrac{3}{4}$$

and
$$p(x, z) < p(y, z).$$

Thus, (−7c) is clearly not absurd.

To return then, to the 1954 argument, Popper's condition of adequacy (7) requires interpretation before it can be applied and it admits of (legitimate) interpretations that are incompatible with the interpretation Popper assumes (7a) (i.e., the denial of (−7a)). If (7) *is* interpretated as (7a) then Popper's 1954 argument has the trivial consequence described by Kemeny. Instead of obtaining

(3c) $\qquad\qquad Cc(xy, z) > Cc(x, z)$

from (7) (i.e., (7a)) we have only

(3p) $\qquad\qquad Cp(xy, z) > Cp(x, z)$

which proves what no one ever doubted, namely, that $Cp \neq p$, i.e., that the explicatum of '*difference* in probability' is not an adequate explicatum for 'probability.' Again, to be able to *derive* a formula that is incompatible with the false

(11) $\qquad\qquad p(xy, z) > p(x, z)$

Popper *must* assume the interpretation (7a);[44] but granting that assumption, he may only derive (3p). But with (3p) he can only prove the obvious, $Cp \neq p$.

11. REFUTATION OF THE 1956 AND 1959 ARGUMENTS

As soon as we point out that (7) is true only if we take it as (7a) ((7b) is irrelevant here) the spuriousness of the "proof" offered in 1956 (Sec. 5) becomes apparent. With

(7a) $\qquad (Co(y, z) \cdot -Co(x, z)) \supset (Cp(y, z) > Cp(x, z))$

transposed and Cc in the consequent of (17), we cannot obtain (19); and with Cp in the consequent of (17) and (7a), we can only derive what we already know, namely, that $Cp \neq p$. Thus, the 1956 argument must be rejected.

The refutation of the alleged contradiction (Sec. 7) is also easily constructed. To show that Carnap's system is self-contradictory, we must use *Carnap's explicata*. That means we must substitute Cc for p in (24), which gives us (−7c) again. To get the self-contradictory (−7a) we must substitute Cp for p, which only proves again what we already know, namely, $Cp \neq p$.

---

[44] That is, given the definition of "Co(x,z)," he has no other choice.

So far as the quotation from Carnap regarding the self-contradictoriness of (7) is concerned, we can only say that if "Warm" is ambiguous as "confirmed" is, his remark must be false. But this does not have any effect on the present arguments or Carnap's system, so he could grant the falsity of his remark without disturbing these arguments or his system.

CHAPTER IV

BAR-HILLEL'S "COMMENTS" AND UNRESTRICTED UNIVERSALS

1. INTRODUCTION

In this chapter Popper's primary concern seems to change from what was originally a defense of his own theory of confirmation to simply an attack on Carnap's theory. Since the two theories have different explicanda, they are not exactly competitors. However, if Popper could show that Carnap's theory contains some inescapable serious flaw, that in itself would be an important contribution.

I begin by summarizing Bar-Hillel's comments on Popper's 1954 paper.[1] In Section three I show that the two replies Popper offers Bar-Hillel are faulty. Then I construct a reply to a third objection of Bar-Hillel which seems to be effective. In Sections four and five some technical terms are introduced. These are necessary in order to understand the theorems and arguments in the remaining sections of this chapter and Chapter V. In Section six I present Carnap's theorems for the degree of confirmation of unrestricted universals[2] and Popper's criticism of them. The discussion of these theorems in the last two sections involves only the first part of what I shall refer to as the *problem of unrestricted universals*. The two parts of this problem may be described briefly as the task of showing that (i) a zero degree of confirmation for unrestricted universals in a universe with an infinite number of individuals is counter-intuitive; and (ii) that the theorems of "instance" and "qualified-instance confirmation"[3] are unsatisfactory. The problem of unrestricted universals is the *third* main issue in the dispute between Popper and Carnap. In the last section of this chapter I reach the somewhat disappointing conclusion that if (i) is

---

[1] These are taken from Bar-Hillel, "Comments on 'Degree of Confirmation' by Professor K. R. Popper," *loc. cit.*, pp. 155–157.
[2] This term is explained in Section 4.
[3] These terms are explained in Chapter V.

taken as a particular (as opposed to a universal) empirical claim then, although it is true, it is fairly uninteresting; and if (i) is taken as a logical claim (i.e., as either logically true or logically false) then, although it might be interesting; its truth-value is still unkwnon.

2. BAR-HILLEL'S 1955 'COMMENTS'

In his 1955 note,[4] Bar-Hillel makes three major points regarding Popper's 1954 paper. I shall summarize these points in the following paragraphs.

A

Bar-Hillel claims first, that the issue between Popper and Carnap "is a verbal one" and he suggests that the apparent dispute might be cleared up by applying this dictionary.[5]

Popper's Terms		*Carnap's Terms*	
1p	absolute (logical) probability	1c	initial confirmation
2p	relative (logical) probability	2c	degree of confirmation
3p	degree of confirmation	3c	relevance measure
4p	supports	4c	is positively relevant to
5p	undermines	5c	is negatively relevant to
6p	independence	6c	irrelevance

We have already seen that 1p and 2p are the explicanda of 1c and 2c respectively.[6] Using our more precise notation, it is clear then, that $Cc(x, z) = p(x, z)$ and $Cc(x) = p(x)$. By comparing Carnap's definitions of 'positive' and 'negative relevance' with Popper's definitions of the classificatory concepts of 'supports' or 'confirms,' and 'undermines' or 'disconfirms,' we may easily prove the accuracy of lines four to six. Popper's definitions are already familiar.[7] Here are Carnap's definitions.

   a. i is *positively relevant* or, briefly, *positive* to
      h ... $\underset{df}{=} C(h, i) > C(h)$.
   b. i is *negatively relevant* or, briefly, *negative* to
      h ... $\underset{df}{=} C(h, i) < C(h)$

---

[4] Bar-Hillel, "Comments on 'Degree of Confirmation' by Professor K. R. Popper," *loc. cit.*, pp. 155–157.
[5] *Ibid.*, p. 155.
[6] See above Chapter III, Section 4.
[7] *Ibid.*, Section 3.

c. i is *relevant* to h ... $\underset{df}{=}$ i is either positively relevant or negatively relevant to h ...

d. i is *irrelevant* to h ... $\underset{df}{=}$ either (1) C(h, i) = C(h), or (2) i is L-false.[8]

The explicandum of these concepts is clearly '*difference* in probability' and not just 'probability.' Thus, the explicandum of these concepts and 4p–6p is obviously the same. 3p is the *explicatum* for the quantitative concept of 'degree of *difference* in probability' which is its explicandum. On the other hand, although Carnap's relevance measure 3c has not been explicated *in its quantitative form*,[9] it is apparent that the *explicandum* of this concept is exactly the same as the explicandum of 3p, namely, 'degree of difference in probability.' These are Bar-Hillel's points regarding the dictionary.

B

Bar-Hillel's second point is that "Popper's point has been fully anticipated and adequately been taken care of by" Carnap.[10] Carnap introduces the following case.[11] We have a chess tournament with ten participants. There are five men m and five women w; and there are juniors and seniors as well as local and foreign players. In tabular form, the players are distributed thus.

	h Local	Foreign
k Junior	mww	mm
Senior	mm	www

Let e be our original information regarding the total number and distribution of players, i.e., the table above summarizes the information given by e. h is 'a local player wins' and k is 'a junior wins.'

---

[8] Carnap, *The Logical Foundations of Probability*, p. 348. When i is L-false (i.e., self-contradictory), c(h,i) is undefined according to D55–3, p. 295, *ibid*.

[9] Chapter VI of Carnap's *The Logical Foundations of Probability* contains an explication of 3c in its classificatory form.

[10] Bar-Hillel, "Comments on 'Degree of Confirmation' by Professor K. R. Popper," *loc. cit.*, p. 155.

[11] Carnap, *The Logical Foundations of Probability*, p. 394, case 3b.

The additional evidence we receive, i, says 'a woman wins.' Thus, we have the following confirmation values.

$$C(h, e) = \tfrac{1}{2} = C(k, e)$$
$$C(h, ei) = \tfrac{2}{5} = C(k, ei) = C(hk, ei)$$

and

$$C(hk, e) = \tfrac{3}{10}.[12]$$

Hence,

$$C(h, ei) < C(h, e) \text{ and } C(k, ei) < C(k, e).$$

So, i is negatively relevant to (undermines) h and k according to b above. But

$$C(hk, ei) > C(hk, e).$$

So, i is positively relevant to (supports) the conjunction hk according to a. Thus, we have a case similar to Popper's since i undermines h and k but supports hk.[13]

Bar-Hillel's comment on this case is this.

> It is far from obvious why 'we should have to say in such a case that i confirms h.k to a higher degree than it confirms either h or k,' as claimed by Popper, rather than describe the situation in the way done in the preceding sentence [i.e., as we did above] or say, in Carnap's terms, that in such a case (the evidence) i is negatively relevant to each of (the two hypotheses) h and k and nevertheless positively relevant to their conjunction...[14]

We might add that another fairly unambiguous way to express the same situation is the following, using only the term 'probability.'

> The probability of h on i is lower than the (initial) probability of h and the probability of k on i is lower than the (initial) probability of k, but the probability of hk on i is higher than the (initial) probability of hk.

C

Bar-Hillel's final point concerns a version of the description argument[15] that Popper included in his 1954 paper. There is, Bar-Hillel claims, no reason why we cannot have both a high degree of confirmation (in the sense of probability) and a high degree of empirical

---

[12] To simplify the notation I have juxtaposed letters to indicate conjunction, instead of inserting a dot between them.
[13] See again Chapter III, Section 2.
[14] Bar-Hillel, "Comments on 'Degree of Confirmation' by Professor K. R. Popper," *loc. cit.*, p. 156.
[15] See Chapter II, Section 3.

content or descriptive power. It is possible to have both by taking advantage of the distinction between initial and relative probability. He claims then, that

> "a 'good' hypothesis is one that has high *initial* (or absolute) informative content, hence a low *initial* confirmation, in Carnap's sense, but a high degree of confirmation, in Carnap's sense, relative to the total available evidence, hence – and this sounds admittedly slightly paradoxical – a low informative content, relative to this evidence.[16]

3. POPPER'S RESPONSE

In this section I shall show that in his rejoinder to Bar-Hillel,[17] Popper unsuccessfully replies to point A, seems to miss the most critical aspect of B and ignores C. In lieu of Popper's own reply to C, I shall construct my own reply to C.

Reply to A

Here is how Popper replies to A.

> Dr. Bar-Hillel's dictionary is incorrect. As to its third line, ... what I call "degree of confirmation" is not the same as Carnap's "relevance measure." *Carnap's relevance measure approaches zero with increasing content of the hypothesis... My "degree of confirmation" approaches one*... Carnap leaves no doubt that from the point of view of his theory, any nonadditive confirmation function is "entirely unacceptable" ...; but my "degree of confirmation" is nonadditive.[18] (Italics added)

You recall that Bar-Hillel's point was that 3p and 3c have the same *explicandum*, namely, '*difference* in probability.' But if you notice the italicized remarks, you will find that Popper is arguing that the third line of the dictionary is "incorrect" because the *explicata* are different. Popper even points out that the differences are "mathematical consequences of the two definitions," which clearly indicates that he is concerned with *explicata* and has completely missed Bar-Hillel's point. His remark following the italicized sentence above as well as his comments on lines four to six are also directed to the explicata. Thus, it is fair to say that Popper has simply failed to understand and satisfactorily reply to point A. But taking A and our discussion in

---

[16] Bar-Hillel, "Comments on 'Degree of Confirmation' by Professor K. R. Popper," *loc. cit.*, p. 156.
[17] Popper, "'Content' and 'Degree of Confirmation': A Reply to Dr. Bar-Hillel," *loc. cit.*, pp. 157–163.
[18] *Ibid.*, p. 162.

Chapter III, Section 4 together, it is apparent that there is no way for Popper to avoid the charge that he and Carnap are concerned with two different explicanda anyhow.

Reply to B

Popper grants that

> Carnap has anticipated, in essence, my [Popper's] examples; he has even drawn from them the conclusion that certain classificatory confirmation concepts are inadequate. But he has not drawn from his examples the conclusion which I have drawn: that all probability functions are inadequate to serve as degree of confirmation.[19]

But then he moves on to an argument that finally results in the 1956 (faulty) derivation of an invalid probability theorem.[20] He never answers Bar-Hillel's important question, namely: Why should one prefer Popper's description of the case rather than Carnap's or Bar-Hillel's (or one using only the term 'probability')? Since each of the suggested descriptions is accurate and not at all absurd according to the individual definitions, there *is* no compelling reason to prefer only one of the descriptions. This is probably why Popper failed to reply to Bar-Hillel's question.

Reply to C

It seems to me that Popper would consider Bar-Hillel's solution of the problem introduced by the description argument extremely *ad hoc*. Why should only a high *initial* empirical content be desirable and *not* a high *relative* empirical content as well? Or, on the other hand, why should only a high *relative* degree of confirmation *or* probability be desirable and *not* a high *initial* degree of confirmation or probability as well? Bar-Hillel is claiming that, say, a hypothesis H is "good" if $Ct(H)$ is high *and* $C(H, e)$ (or $p(H, e)$) is high. But $C(H, e)$ is high only if $Ct(H, e)$ is low. Thus, for Bar-Hillel, a *high initial* empirical content is desirable but a *low relative* empirical content is also desirable. And the question Popper would and I *do* raise is this: Why is a *high relative* empirical content *not* desirable? Bar-Hillel is

---

[19] *Ibid.*, p. 160. Strictly speaking, the conclusion Carnap draws on p. 475, *The Logical Foundations of Probability* is not "that certain classificatory confirmation concepts are inadequate" but that Hempel apparently has two explicanda in mind in C. G. Hempel, "Studies in the Logic of Confirmation," *Mind*, LIV (1945), 1–26, 97–121 and "A Purely Syntactical Definition of Confirmation," *Journal of Symbolic Logic*. VIII (1943), 122–143.

[20] See Chapter III, Section 5, 6 and 11.

evidently interested in having hypotheses with high empirical content. But why should this interest *not* apply to *relative* as well as *initial* empirical content? Surely any reason offered for preferring high empirical content to, say, high degree of confirmation is a reason for preferring both high initial and relative empirical content to high degree of confirmation. This is merely a roundabout way of reasserting my view that a tension exists in the idea of *prudential growth* introduced earlier in the final paragraph of Chapter II. The truth of the matter seems to be that what is required is not a question of cutting the pie as it were between Popper's and Carnap's insights, but of delicately fusing those insights into a single concept.

4. RESTRICTED AND UNRESTRICTED UNIVERSAL AND EXISTENTIAL SENTENCES

In this section and the next one I shall briefly introduce a few technical terms. The terms introduced in this section are fairly common. Those discussed in section five are peculiar to Carnap's system.

An *unrestricted simple law* is, for Carnap, a sentence that "speaks in a purely general way about the individuals of the system in question without referring to any particular individual," e.g., "All swans are white."[21] *Restricted simple laws* apply "to all individuals with the exclusion of some specified individuals," e.g., "With the exception of a certain kind of swan found only in Australia, all swans are white."[22] Carnap's 'unrestricted simple law' corresponds to Popper's 'strictly universal statement.' A *strictly universal statement* is, for Popper, a statement that "claims to be true for any place and any time," e.g., "All ravens are black."[23] It is contrasted with *numerically universal statements* which refer "only to a finite class of specified elements within a finite individual (or particular) spatio-temporal region," e.g., "Of all human beings now living on the earth it is true that their height never exceeds a certain amount (say 8 ft.)."[24] We shall only be concerned with unrestricted simple laws or strictly universal sentences, and shall refer to them hereafter as *unrestricted universals*.

The contradictory of an unrestricted universal is an unrestricted existential statement, e.g., 'Some crows are not black.' Because "what

---

[21] Carnap, *The Logical Foundations of Probability*, p. 142: the example is Carnap's.
[22] *Ibid.*
[23] Popper, *The Logic of Scientific Discovery*, p. 62: the example is Popper's.
[24] *Ibid.*

we call natural laws, have the logical form of strictly universal statements ... [Popper claims] they can be expressed in the form of negations of strictly existential statements or ... *non-existence statements* ...."[25] And Carnap has indicated his acceptance of Popper's view.[26]

It is generally claimed[27] that both kinds of universals may be falsified. Unrestricted universals may not be verified if either (1) singular sentences are not verifiable, or (2) an infinite number of instances must be examined. If (1) is denied then (some, at least) restricted universals might be verified. But this cannot be said for unrestricted universals. Even if (1) is denied, if (2) is granted then unrestricted universals may not be verified.[28] That both kinds of universals may be falsified *if* singular sentences are verifiable may be proven by recalling the definition of contradictory sentences, i.e., one must be true and the other false. Thus, if we are certain that a particular proposition is true then we may be certain that its universal contradictory is false.

On the other hand, it is generally claimed that both kinds of existential sentences (i.e., unrestricted and restricted) may be verified (if *any* sentences are verifiable) but unrestricted existential sentences may not be falsified. Because unrestricted existential sentences apply to an infinite number of objects, the examination of all its instances is impossible; thus, complete falsification of such sentences is impossible.

5. Q-PREDICATES AND LOGICAL WIDTH

In order to understand the theorems that are particularly relevant to unrestricted universals, we must explain Carnap's two technical concepts, "Q-predicate" and "logical width." Consider a language with only two undefined or primitive predicates P and R, which represent only properties, not relations. Let 'pi' stand for the number

---

[25] *Ibid.*, p. 69.
[26] Carnap, *The Logical Foundations of Probability*, p. 406.
[27] The views expressed in this paragraph and the next may be found in Hempel, "Studies in the Logic of Confirmation," *loc. cit.*, pp. 112–113; Henry K. Mehlberg, *The Reach of Science* (Toronto: University of Toronto Press, 1958), pp. 254–255. An interesting critique of a view contrary to that expressed in the first sentence of our text may be found in Adolf Grunbaum, "The Falsifiability of Theories: Total or Partial? A Contemporary Evaluation of the Duhem-Quine Thesis," *Boston Studies in the Philosophy of Science*, ed. Marx W. Wartofsky (Dordrecht-Holland: D. Reidel Publishing Co., 1963), pp. 178–195.
[28] Carnap, of course asserts (1) and (2). See, for example, Carnap, "Testability and Meaning," *loc. cit.*, p. 425. Norman Malcolm tries to prove that (1) and (2) are false in "The Verification Argument," *Philosophical Analysis: A Collection of Essays*, ed. Max Black (Englewood Cliffs, N.J.: Prentice-Hall, Inc., 1963), pp. 229–279.

of these predicates; i.e., pi = 2.[29] If N is the number of individual constants $a_1, a_2, \ldots a_N$ in our simple language, we may use, following Carnap, 'L_N^{pi}' to represent that language; i.e., L_N^{pi} represents a language with pi primitive predicates and N individual constants. Carnap calls predicates that are either primitive or "constructed from [primitive] predicates with the help of connectives," "molecular predicates." "Q-predicates" are molecular predicates that designate only properties (i.e., not relations) and the properties designated are "Q-properties."[30] A "Q-predicate expression" is a conjunction of all primitive predicates (or the negations of some or all of them) in lexicographical order.[31] Q-predicates are simply abbreviations of Q-predicate expressions. Figure 1 shows the four possible Q-predicates for L_N^{pi}.[32]

Q-predicate expression	Q-predicate
P · R	Q1
P · —R	Q2
—P · R	Q3
—P · —R	Q4

Fig. 1. – Q-predicates for L_N^{pi}.

The properties for which Q-predicates stand can not be "subdivided into several factual properties." Thus, they represent the *infimae species* of Aristotelian logic.[33] The number of Q-predicates in L^{pi} is $k = 2^{pi} = 4$. k is one of the two most important variables in the theorems to be introduced in the next section and in the next chapter.[34]

Now, in Carnap's words

The Q-properties of a [language] system L are the strongest non-L-empty properties expressible in L. If M is any non-L-empty molecular property in L,

---

[29] In languages more complicated than the one considered here, pi is disregarded. See Carnap, "Replies and Systematic Expositions," p. 974.

[30] Carnap, *The Logical Foundations of Probability*, pp. 124-125.

[31] *Ibid.*, Carnap's technical definition of "lexicographical order" may be found *ibid.*, p. 68. For our purposes, alphabetical order would do.

[32] Notice that the number of individual constants N is irrelevant for determining Q-predicates.

[33] *Ibid.*, p. 126.

[34] k has a different meaning in more complicated languages and although none of the theorems or arguments we shall consider involve complicated languages, it is worthwhile to explain the other meaning of k. In more complicated languages, predicates are classified into families, e.g., we might have a family of colors and another of odors. In such languages k stands for the number of predicates in any given family. Thus, if P and R belonged to two different families in L^2, then Carnap now would say that L^2 is a language with two families, each having k = 2 predicates (viz., P and -P, and R and -R). See Carnap, "Replies and Systematic Expositions," *loc. cit.*, pp. 973-974.

then M is uniquely analyzable into a disjunction of Q's or one Q. Let w be the number of these Q's. w is called the (logical) width of M; if M is L-empty, we assign to it the width w = 0. If w is the width of M, w/k is its relative width.[35]

The width of each Q-predicate in Figure 1 is (by convention) w = 1 and the relative width of each of these predicates is w/k = $\frac{1}{4}$. w is the second of the two most important variables in the theorems to be introduced.

6. THE DEGREE OF CONFIRMATION OF UNRESTRICTED UNIVERSALS

According to Carnap's definition of 'degree of confirmation,' the degree of confirmation of unrestricted universals approaches zero as the number of individuals in the universe approaches infinity. We may prove this by considering the following two theorems.[36] The first applies to cases in which no unfavorable evidence has been found. The second applies to cases in which unfavorable evidence has been found.

A

Consider again our simple language with $pi = 2$ primitive predicates P and R, and $N = \infty$ individuals $a_1, a_2, ..., a_\infty$. We are interested in the degree of confirmation of the unrestricted universal U 'All P is R' i.e., '$(x)(Px \supset Rx)$.' Let M stand for '$P \cdot -R$' which has a width $w_1 = 1$.[37] Then U asserts that 'There are no M' and has a strength $w_1 = 1$.[38] If e is a sample of s different individuals then the degree of confirmation of U given e is

(A1) $$c^*(U, e) = \frac{\binom{s + k - 1}{w_1}}{\binom{N + k - 1}{w_1}}.$$

---

[35] Carnap, *The Continuum of Inductive Methods*, pp. 10–11. A longer discussion of the concept of "logical width" may be found in Carnap, *The Logical Foundations of Probability*, pp. 126–130. A property is "L-empty" if no individual has that property.
[36] Our discussion follows Carnap, *ibid.*, pp. 570–571.
[37] By convention, each Q-predicate has the same width, namely, one. *Ibid.*, p. 127.
[38] *Ibid.*, p. 144, T37–3a.

The right side of this equation is an abbreviation of

(A2)
$$\frac{\dfrac{(s + k - 1)!}{w_1!((s + k - 1) - w_1)!}}{\dfrac{(N + k - 1)!}{w_1!((N + k - 1) - w_1)!}}$$

which may be obtained from the familiar law for the number of possible combinations of m things taken n at a time

$$\binom{m}{n} =_{df} \frac{m!}{n!(m - n)!}$$

by substituting w_1 for n, and $s + k - 1$ for m to get the numerator of (A2) and $N + k - 1$ for m to get the denominator. When $n = 1$, $\binom{m}{n} = m$.[39] This simplifies our explanation of (A1). With $w_1 = 1$ the right side of (A1) becomes a fraction with infinity as its denominator (because $N = \infty$).[40] Thus, $c^*(U, e) = 0$ when $N = \infty$, and it is easy to see that $c^*(U, e)$ *approaches* zero as N approaches infinity.[41]

B

Now consider a case in which all things are as in A but our new evidence e' says that s_1 of our sample of s individuals violates U. Because our new evidence e' is incompatible with U (i.e., in Carnap's terminology e' and U are "L-exclusive") "in this case there is no point in taking as hypothesis the law in its original form U .... We take instead the corresponding restricted law U' which says that all individuals not belonging to the sample described in e' have the property PR ('all unobserved P is R')."[42] We have then

(B)
$$c^*(U', e') = \frac{\binom{s + k - 1}{s_1 + w_1}}{\binom{N + k - 1}{s_1 + w_1}}$$

---

[39] *Ibid.*, pp. 150–151.

[40] e and therefore s, of course, will always be finite because they represent some observation or set of observations.

[41] In "Replies and Systematic Expositions," *loc. cit.*, p. 977, Carnap writes, "So far as I know, no other author has given a satisfactory system of logical probability in which [the following] does not hold. Let U be a universal factual sentence for an infinite domain of individuals. Let e be an arbitrary sentence without quantifiers. Then C(U,e) = O." Thus, according to Carnap, systems with theorems yielding C(U,e) = O (i.e., p(U,e) = O) are more satisfactory than systems without this result. In Chapter VIII we shall examine a system constructed by Jaakko Hintikka that does not have this result, i.e., it allows unrestricted universals to have positive probability values.

[42] Carnap, *The Logical Foundations of Probability*, p. 571.

It may readily be seen that as long as $N = \infty$, $c^*(U', e') = 0$ according to (B); and as N approaches infinity, $c^*(U', e')$ approaches zero.

Popper claims that Carnap's definition of degree of confirmation is inadequate because "it attributes to the best confirmed laws such as 'sugar is soluble in water' precisely the same 'degree of confirmation' as to laws which are always refuted (or which are self-contradictory)."[43] And he tries to show that "Carnap, by his own standards, should have discarded his definition of degree of confirmation as *inadequate* ...."[44] Popper's argument runs as follows. Carnap claims that a definition of 'degree of confirmation' is adequate if it yields confirmation values that are "sufficiently in agreement with our intuitive values."[45] Now it happens that many physical laws are regarded "as 'very reliable,' 'well founded,' 'amply confirmed by numerous experiences.' "[46] But these laws (when they are unrestricted universals) can only have a degree of confirmation equal to zero according to Carnap's definition (by (A1) and (B)). Thus, the confirmation-values assigned according to Carnap's definition are practically complementary to those assigned by our intuition.[47] And that means, according to Carnap's criterion, that the definition should be discarded.

7. THE REPLIES OF CARNAP AND BAR-HILLEL

Carnap's reply to the above challenge is quite brief. He denies that a confirmation-value of zero for an unrestricted universal in a universe with an infinite number of individuals is counter-intuitive. He writes

> On the contrary, my whole discussion in (§ 110G) tries to show that the result, in spite of the first appearance, is not counter-intuitive. I can easily imagine that a reader might remain unconvinced by my arguments. But it is thoroughly puzzling to me how any reader could have the impression that I myself believed the proposition which I tried so hard to refute.[48]

---

[43] Popper, "'Content and 'Degree of Confirmation': A Reply to Dr. Bar-Hillel," *loc. cit.* pp. 158–159. This objection has been raised by a number of writers. See, for example, Donald C. Williams, "Professor Carnap's Philosophy of Probability," *Philosophy and Phenomenological Research*, XIII (1952), p. 117; Ernest Nagel, "Carnap's Theory of Induction," in Schilpp, *op. cit.*, p. 799; Barker, *op. cit.*, p. 87; Mehlberg, *op. cit.*, pp. 198–199.
[44] Popper, "'Content' and 'Degree of Confirmation': A Reply to Dr. Bar-Hillel," *loc. cit.*, p. 159.
[45] Carnap, *The Logical Foundations of Probability*, p. 232.
[46] *Ibid.*, p. 571.
[47] In Popper's words, "The gap between the intuitive value for a very well confirmed law and the value zero just could not be wider." Popper, "'Content,' and 'Degree of Confirmation': A Reply to Dr. Bar-Hillel," *loc. cit.*, p. 159.
[48] R. Carnap, "Remarks on Popper's Note on Content and Degree of Confirmation," *B.J.P.S.*, VII (1956), p. 244.

By the "proposition which I tried so hard to refute," I presume Carnap means this one (i): a confirmation-value of zero for an unrestricted universal in a universe with an infinite number of individuals is counter-intuitive. Now, in § 110 G, (i) is challenged only indirectly (by implication). The points Carnap makes explicitly in that section are first, that engineers are not concerned with unrestricted universals, and second, that such sentences are not necessary for ordinary people or scientists. On the first point, he writes,

> When [an engineer] says that a law is very reliable, he does not mean to say that he is willing to bet that among the billion of billions, or an infinite number, of instances to which the law applied there is not one counter-instance, but merely that this bridge will not be a counter-instance, or that among all bridges which he will construct during his lifetime there will be no counter-instance.[49]

This sentence seems to imply that (i) is false because engineers, at least, are not interested in the confirmation-value of unrestricted universals and therefore, *any*[50] value one assigns to these sentences will not be counter-intuitive (to engineers) because engineers have no intuition at all about these values. This is, of course, a rather dubious claim.[51] But the important point for us is that the remarks in *The Logical Foundations of Probability do* seem to support Carnap's 1956 claim.[52] That is, Popper's claim that Carnap believed or believes (i) is true, seems to be *false*.

On the other hand, Bar-Hillel claims that we must distinguish "guided" from "unguided intuition," and change Carnap's "agreement with intuition" criterion to an "agreement with guided intuition" criterion.[53] Evidently, according to Bar-Hillel, a zero degree of confirmation for unrestricted universals in a universe with an infinite number of individuals is not then, "counter-*guided*-intuitive." He grants however, that "it must be admitted that the qualification 'guided' that has to be prefixed to 'intuition' in the phrase 'sufficient agreement with intuition' is a major change in the formulation of the adequacy conditions, and a change to the better, to my mind."[54]

---

[49] Carnap, *The Logical Foundations of Probability*, p. 572.
[50] That is, the confirmation-value of every unrestricted universal could be 1 or 1/2 and still not be counter-intuitive.
[51] It will be challenged in the next section.
[52] I am assuming that Carnap wants us to understand that his intuition is more or less in agreement with that of the engineers.
[53] Bar-Hillel, "Further Comments on Probability and Confirmation: A Rejoinder to Professor Popper," *loc. cit.*, p. 247.
[54] *Ibid.*

Thus, from Bar-Hillel's point of view, (i) is *true*, but may easily be falsified by altering the adequacy criterion.

8. FINAL REMARKS ON THE DEGREE OF CONFIRMATION OF UNRESTRICTED UNIVERSALS

As I suggested above, Carnap's (implied) claim that engineers have no intuition about or interest in unrestricted universals seems doubtful. I certainly have made no statistical investigation of what engineers "mean to say," but then, Carnap probably never has either. It seems very likely though, that some engineer somewhere has some intuition about or interest in unrestricted universals. But even if every engineer agrees with Carnap, some pure scientists and some philosophers of science certainly disagree with him.[55] So it seems that the most he could prove with his claim is that some people (besides Carnap) think (i) is false. But what, after all, have Popper and those who agree with him proved? Nothing more than this: some people think (i) is true. So far as I can tell, none of those objectors referred to in Section six (namely, Williams, Nagel, Barker and Mehlberg) are confused about Carnap's *explicandum*. Thus, though it might be claimed that Popper thinks (i) is true because he is confused about the explicandum of Carnap's 'degree of confirmation,' this claim cannot be made for the other objectors. But again, what follows from that? Roughly the same thing: some people (who are not confused) think (i) is true.

We have come, I think, a fairly long way to reach a disappointing conclusion. The "arguments" of both Popper and Carnap (and Bar-Hillel) seem to be nothing more than assertions about what some people think about the truth-value of (i). And as we might have expected, these assertions have persuaded neither Popper nor Carnap to alter his position regarding (i). But again, considering (i), we should have expected an inconclusive result because (i) is an extremely vague claim. It is evidently not a *universal* empirical claim (i.e., the claim that everyone finds a zero probability value for unrestricted universals counter-intuitive), because all the disputants know such a claim would be false. However, (i) is, as we have just seen, very uninteresting if it is taken as a *particular* empirical claim. What, after all, is so significant about the fact that some people think a zero probability value for such sentences is counter-intuitive and some people do not

---

[55] See Section 6.

think so? Hence, assuming that (i) is either an empirical claim, logically false or logically true, we have two interesting possibilities left. First, (i) is logically true; and second, (i) is self-contradictory. None of the disputants tried to prove either possibility and it is not difficult to see why. To prove that either possibility is the case, one would have to explicate the concept of intuition and show that a zero probability value for unrestricted universals is logically compatible or, alternatively, logically incompatible with this explicatum. This, of course, suggests an extremely arduous investigation. But more important than that, I think, is the fact that it is by no means obvious that such an explicatum would have *any* relation to a zero probability value for unrestricted universals. At any rate, it is not my intention to attempt such an investigation and there is no explicit evidence that the disputants believe (i) to be a logical claim anyhow. It seems more likely that the disputants intend (i) as an empirical claim, because none of them ever mentions the fact that it *would* require a considerable effort (much less does anyone show signs of *expending* such an effort) to *prove* it is *either* logically false *or* logically true. Therefore, it seems likely that (i) is at best only a *particular* empirical claim and a fairly uninteresting one at that. I suggest then, that we simply abandon this part of the problem of unrestricted universals. This is not to say that we shall abandon the whole problem. We have not, in fact, considered Carnap's complete view.

After claiming that engineers, at least, are not interested in the degree of confirmation of unrestricted universals, Carnap introduces the concepts of "instance confirmation" and "qualified-instance confirmation." Then he claims that there is *no need* for a high degree of confirmation for unrestricted universals because a high degree of instance or qualified instance confirmation is all that is required "for practical purposes of everyday life or for theoretical purposes of science."[56] Thus, if Popper is going to prove (to Carnap) that Carnap's definition of 'degree of confirmation' is inadequate because Carnap's treatment of unrestricted universals is unsatisfactory, Popper must prove (ii) that the theorems for instance and qualified-instance confirmation are unsatisfactory. If these theorems are unsatisfactory then Carnap would have to admit that he has no satisfactory solution for the problem of unrestricted universals. And that, I should think, would be a serious objection to Carnap's theory.

[56] Carnap, *The Logical Foundations of Probability*, pp. 572–574.

CHAPTER V

INSTANCE AND QUALIFIED-INSTANCE CONFIRMATION

1. INTRODUCTION

In this chapter I shall show that (ii) (i.e., the theorems of "instance" and "qualified-instance confirmation" are unsatisfactory) is true. The *third* issue of the dispute between Popper and Carnap is Popper's claim that Carnap's definition of 'degree of confirmation' is inadequate because Carnap's treatment of unrestricted universals is unsatisfactory. The defense of (ii) then, is a defense of Popper's view of this part of the dispute. If the objections presented in sections four, five and seven are sound, the theorems of instance and qualified-instance confirmation must be rejected. That would mean Carnap has *not* offered a satisfactory treatment of unrestricted universals.

I begin by introducing the theorems of instance and qualified-instance confirmation. In section three, Popper's first objection to these theorems is refuted. In section four, a brief comment of Popper's regarding the theorem of instance confirmation is developed into an apparently inescapable objection to this theorem. In the next two sections two paradoxes are introduced, namely, one I refer to as "Popper's paradox," and Hempel's paradox. Popper's paradox is not the paradox made famous by C. G. Hempel,[1] and it does not admit of the solution Hempel applies to Hempel's paradox. Popper's paradox, in fact, destroys the theorem of qualified-instance confirmation by showing that the theorem leads to paradoxical results. In section six, Hempel's paradox is introduced and resolved following Hempel's own procedure. In the last section, an objection to both theorems is introduced that seems to be inescapable only for the theorem of instance confirmation.

---

[1] Hempel, "Studies in the Logic of Confirmation," *loc. cit.*, pp. 1–26, 97–121.

2. THE THEOREMS OF INSTANCE AND QUALIFIED-INSTANCE CONFIRMATION

Let U be a simple law of the form '$(x)(Mx \supset M'x)$' (e.g., 'all swans are are white'). By the *instance confirmation* of U on some evidence e Carnap means "the degree of confirmation, on the evidence e, of the hypothesis that a new individual not mentioned in e fulfills the law U."[2] Let h be that hypothesis (i.e., h might be '$Mb \supset M'b$' provided that b is not mentioned in e). Then in symbols we have

$$c_i^*(U, e) \underset{df}{=} c^*(h, e)$$

"By the *qualified-instance confirmation* of the law that all swans are white we mean the degree of confirmation for the hypothesis h' that the next swan to be observed will likewise be white"[3] (i.e., h' might be '$M'b$' provided that we have already observed Mb (j for short) and b is not mentioned in e). Then in symbols we have

$$c_{qi}^*(M, M', e) \underset{df}{=} c^*(h', ej)$$

We cannot make Carnap's summary of the two theorems more concise.

Let U be '$(x)(Mx \supset M'x)$.' Let 'M_1' be defined by '$M.-M'$' (Non-white swan) and 'M_2' by '$M.M'$' (White swan). Let the widths of 'M_1' and 'M_2' be w_1 and w_2, respectively. Let e be a report about s observed individuals saying that s_1 of them are M_1 (negative cases) and s_2 are M_2, while the remaining ones are $-M$ (Non-swan) and hence neither M_1 nor M_2. Then the following holds:

(1) $$c_i^*(U, e) = 1 - \frac{s_1 + w_1}{s + k}$$

(2) $$c_{qi}^*(M, M', e) = 1 - \frac{s_1 + w_1}{s_1 + w_1 + s_2 + w_2}.\text{[4]}$$

Since N, the number of individuals in the universe, does not appear in these theorems, the fact that we might have $N = \infty$ presents no problem. Carnap claims that because

It can be shown that, if the number s_1 of observed negative cases is either 0 or a fixed small number, then, with the increase of the sample size s, both c_i^* and c_{qi}^* grow close to 1 ... This justifies the customary manner of speaking of "very reliable" or "well-founded" or "well-confirmed" laws, provided we interpret

---

[2] Carnap, *The Logical Foundations of Probability*, p. 572.
[3] *Ibid.*
[4] *Ibid.*, p. 573.

INSTANCE AND QUALIFIED-INSTANCE CONFIRMATION

these phrase as referring to a high value of either of our two concepts just introduced.[5]

Since Carnap has introduced the term 'reliable,' it is reasonable to ask now: What exactly is the relation between degrees of confirmation and degrees of reliability? The answer is twofold. In the first place, the latter are defined and measured by the former. In this respect the concept of reliability is related to the concept of confirmation as the concepts of positive relevance, negative relevance, classificatory confirmation, etc. are related to the quantitative concept of confirmation. Some function of the quantitative concept of confirmation is used in the definiens of all of these other concepts. In the second place, since the degree of confirmation of an unrestricted universal (law or theory) is always zero (in universes with an infinite number of individuals) but the degree of reliability of such universals is always positive, the two concepts are evidently distinct. Thus, although reliability is defined and measured in terms of confirmation, the concepts are not identical as those in, say, "All bachelors are unmarried males."

3. REFUTATION OF OBJECTION ONE

In 1955[6] Popper offered three objections to (1) and (2) (i.e., the theorems of instance and qualified-instance confirmation). I will present and refute the first of these in this section, and consider the other two in Sections 4 and 5.

Popper's first objection is this. According to (1) and (2), the degree of confirmation of a "regularly refuted" hypothesis might be very high although "we will all say that the hypothesis had been amply refuted by the evidence."[7] For example, Popper suggests the following case. Consider "a universe of coin tosses with only two predicates: 'coming up heads' and 'coming up tails.'" Our hypothesis is "all tosses always come up heads." Let e say that half of 100 tosses have been negative for h, (i.e., tails). Considering 'coming up tails' as simply the negation of 'coming up heads,' we have a universe with one *primitive* predicate H and $k = 2^1 = 2$ Q predicates.[8] $w_1 = 1 = w_2$. Thus,

---

[5] *Ibid.*
[6] Popper, "'Content' and 'Degree of Confirmation': A Reply to Dr. Bar-Hillel," *loc. cit.*, p. 161.
[7] *Ibid.*
[8] See Chapter IV, Section 5.

according to (1) and (2) we have

$$c_i^*(h, e) = 1 - \frac{50 + 1}{100 + 2} = \tfrac{1}{2}$$

$$c_{qi}^*(h, e) = 1 - \frac{50 + 1}{50 + 1 + 50 + 1} = \tfrac{1}{2}$$

However, because a falsified hypothesis should have a degree of confirmation of zero, (1) and (2) are unacceptable according to Popper.

As I have presented Popper's objection, his error is apparent. He claims that the degree of confirmation of h 'all tosses always come up heads' should be, but is not, zero. But the degree of confirmation of h *is* zero. The degree of *instance* and *qualified-instance* confirmation of h is $\tfrac{1}{2}$. h *has been* "amply refuted," and only *instances* or *qualified-instances* of h may have a degree of confirmation greater than zero. In Bar-Hillel's words,

> The truth is, of course, that Carnap too assigns to h a very low confirmation value – for an infinite universe just zero – whereas the confirmation value $\tfrac{1}{2}$ is assigned not to h – and this in spite of the English formulation – but rather to an instance of h.[9]

It is clear then, that Popper's first objection is defective and based on a misunderstanding.

4. OBJECTION TWO

Popper's second objection concerns only (1). He claims that "in a sufficiently complex world, its value will be very close to zero for any complex predicate."[10] His remark is extremely brief,[11] but I shall construct the objection in detail. Consider (1). If s_1/s is very large, w_1/k will have little effect on the final value of (1). Popper is suggesting that we might obtain a similar result by making w_1/k very large. In that case our "empirical factor" s_1/s would have little effect on the final value of (1). His point may be elucidated with the help of an

---

[9] Bar-Hillel, "Further Comments on Probability and Confirmation, A Rejoinder to Professor Popper," *loc. cit.*, p. 248.
[10] Popper, "'Content' and 'Degree of Confirmation': A Reply to Dr. Bar-Hillel, *loc. cit.*, p. 161.
[11] Bar-Hillel's only comment on Popper's remark is, "I did not understand his first reason for the claim of inadequacy of the unqualified instance confirmation." Bar-Hillel, "Further Comments on Probability and Confirmation: A Rejoinder to Professor Popper, *loc. cit.*, p. 247.

INSTANCE AND QUALIFIED-INSTANCE CONFIRMATION 53

example. Consider a language with 12 primitive predicates (representing only properties) $P_1, P_2, \ldots, P_{12}$ and one hundred individuals $a_1, a_2, \ldots, a_{100}$. The hypothesis h we wish to confirm is 'Every individual has only Q_1' when Q_1 is an abbreviation of the Q-predicate expression '$P_1 \cdot P_2 \cdot \ldots \cdot P_{12}$.'[12] Thus, if we find any individual with *any* predicate other than Q_1, we must count this as unfavorable evidence for h. Now, in L_{100}^{12} we have $k = 2^{12} = 4096$. w_1 in (1) is the width of the unfavorable predicate. Since only Q_1 is favorable, the width of the unfavorable predicate is the sum of the widths of $Q_2, Q_3, \ldots, Q_{4096} = 4095 = w_1 = k - 1$.[13] Here then, is an example of Popper's point. Even if our evidence e tells us that all of 99 individuals examined have been favorable for h, according to (1) we have

$$c_i^*(h, e) = 1 - \frac{4095}{99 + 4096} = \frac{20}{839}$$

or approximately .02; and by increasing the number of primitive predicates in our lagnuage, we may lower this value as much as we like. Thus, we have constructed a language and hypothesis such that the confirmation-value yielded by (1) will be close to zero given almost any amount of favorable evidence.[14]

5. OBJECTION THREE: POPPER'S PARADOX

Popper's view of the seriousness of his third objection is ambiguous.[15] His objection is that "the qualified-instance confirmation ... is inconsistent: it is hit by the paradox of confirmation."[16] A few lines later he writes, "The paradox of confirmation is not so serious as it looks; there is a general method of avoiding it by symmetrization." Unfortunately (for Carnap), as is shown below, the objection is *not* eliminated. The paradox Popper has in mind is *not* the paradox made famous by C. G. Hempel.[17] Popper's paradox may be explained thus.

---

[12] That is, no individual lacks any *primitive* predicate.
[13] Carnap, *The Logical Foundations of Probability*, p. 129, T32-5a.
[14] Notice that if we let k stand for the number of predicates in only one family, the objection may still be raised. See Chapter IV, Section 5.
[15] His position regarding the scope of this objection is also ambiguous. He originally claimed that (1) "shares the absurdity of the (rectified) qualified-instance confirmation." "'Content' and 'Degree of Confirmation': A Reply to Dr. Bar-Hillel," *loc. cit.*, p. 161. Later he claimed that he "asserted *only* of the 'qualified-instance confirmation' that it is hit by the paradox of confirmation." Popper, "Adequacy and Consistency" A Second Reply to Dr. Bar-Hillel, *loc. cit.*, p. 251.
[16] Popper, "'Content' and 'Degree of Confirmation': A Reply to Dr. Bar-Hillel, *loc. cit.*, p. 161.
[17] See Section 6.

Suppose we are given the law L_1 '$(x)(Mx \supset Nx)$' whose degree of confirmation is 0. Carnap tells us to consider the degree of confirmation of some qualified-instance of L_1, say, h 'Nb' and let *that* represent the reliability of L_1. On the basis of h then, we say the degree of reliability of L_1 is r. Now suppose we transpose L_1 and call it L_2'$(x)(-Nx \supset -Mx)$.' The degree of confirmation of L_2 is 0; so, again Carnap tells us to consider the degree of confirmation of some qualified-instance of L_2, say, H '$-Mb$' and let *that* represent the reliability of L_2. On the basis of H then, we say the degree of reliability of L_2 is r' $\neq$ r.

Thus, the degree of reliability of our law clearly varies as we vary its linguistic formulation. And, according to Popper

> This fact constitutes precisely the so-called paradox of confirmation: an engineer building a bridge who asks himself what is the reliability of the law L on which he bases his calculations will be deeply disturbed if he is told that *this depends entirely on the choice of words in formulating the law*... If the engineer's knowledge of... logic, or his logical intuition, tells him correctly that these two laws are *only verbally different,* then he will hardly be satisfied by Professor Carnap's assertion that his qualified-instance confirmation "represents still more accurately what is vaguely meant by the reliability of a law."[18] (Italics added.)

Popper's "general method" of avoiding this paradox is, I am sorry to say, not particularly clear to me. I shall, however, reconstruct it and and suggest a reason for doubting its plausibility. To begin with let

(5) $\qquad c^*_{qi}(M, M', e) = c^*(h', ej) = c^*(M'b, e \cdot Mb)$

and

(6) $\qquad c^*_{qi}(-M', -M, e) = c^*(H', eJ) = c^*(-Mb, e \cdot -M'b)$.

Popper claims that if we replace 'h' by 'j $\supset$ h' and 'j' by 'h' $\supset$ j' in (5) the paradox may be avoided. That is, instead of determining the degree of confirmation of 'M'b' given 'e and Mb,' we should determine the degree of confirmation of 'Mb $\supset$ M'b' given 'e and (M'b $\supset$ Mb).' It seems to me that all Popper's claim amounts to is this: If some *other* hypothesis and some *other* evidential statement is used to determine the degree of confirmation of some law L then Popper's paradox may be avoided. But since Carnap's theorem prescribes the use of a specific hypothesis and evidential statement, Popper's method is not even applicable. Carnap's theorem (2) cannot be saved by showing that some other theorem might be acceptable.

---

[18] Popper, "Adequacy and Consistency: A Second Reply to Dr. Bar-Hillel," *loc. cit.*, p. 252.

6. OBJECTION FOUR: HEMPEL'S PARADOX

According to Carnap's remarks in the paragraph preceding (1) and (2), given the law U '(x)(Mx $\supset$ M'x),' we count individuals with the predicate 'MM'' as favorable evidence and individuals with 'M — M'' as unfavorable evidence for instances or qualified-instances of U. Individuals lacking the antecedent M will be irrelevant. We might say then, that sufficient conditions[19] of confirmation and disconfirmation have been implicitly stipulated. Specifically, it is assumed that

(a) Individual b is confirming evidence for U if b has the properties designated by the antecedent and consequent of U.

(b) Individual b is disconfirming evidence for U if b has the property designated by the antecedent but lacks the property designated by the consequent of U.

(c) Individual b is irrelevant to U if b lacks the property designated by the antecedent of U.

In the last section, we saw that one of Carnap's conditions of adequacy for any explicatum of probability is

(d) *L-equivalent hypotheses.* If h and h' are L-equivalent, then C(h, e) = C(h', e).[20]

Granting (a) — (d), we get the paradox made famous by C. G. Hempel.[21]

---

[19] Both Carnap and Hempel have shown that the criterion of confirmation suggested is sufficient only if the hypotheses considered are simple laws with *one* variable. "That in the case of laws with several variables it is not even sufficient is shown by Hempel with the help of the following counter-example [Hempel, "Studies in the Logic of Confirmation," *loc. cit;* p. 13], which is interesting and quite surprising. Let the hypothesis be the law, "(x) (y) (-(Rxy . Ryx)) $\supset$ (Rxy . -Ryx))." ... Now the fact described by "Rab . -Rba" fulfills both the antecedent and the consequent in the law; hence the fact should be taken as a confirming case according to Nicod's criterion [i.e., according to (a) above]. However, since the law stated is L-equivalent to "(x) (y)Rxy," the fact mentioned is actually disconfirming." Carnap, *The Logical Foundations of Probability*, pp. 469–470.

[20] (d) is the quantitative counterpart of Hempel's "Equivalence Condition" for the classificatory concept of confirmation. Both conditions require logically equivalent hypotheses be to affected in the same way by the same evidence. See *ibid.*, p. 474.

[21] Hempel, "Studies in the Logic of Confirmation." *loc. cit.*, pp. 13–21. A number of articles have been written about Hempel's paradox. See, for example; H. G. Alexander, "The Paradoxes of Confirmation: A Reply to Dr. Agassi," *B.J.P.S.*, X (1959), 229–234; J. Agassi, "Corroboration Versus Induction," *B.J.P.S.*, IX (1958), 311–317; C. H. Whiteley, "Hempel's Paradoxes of Confirmation," *Mind*, LIV (1945), 156–158; I. J. Good, "The Paradox of Confirmation," *B.J.P.S.*, II (1960), 145–149; J. W. N. Watkins, "When Are Statements Empirical," *B.J.P.S.*, II (1960), 287–308; J. W. N. Watkins, "Confirmation Without Background Knowledge," *B.J.P.S.*, II (1960), 318–320; J. W. N. Watkins, "A Rejoinder to Professor Hempel's Reply," *Philosophy*, XXXIII (1958), 349–355; J. W. N. Watkins, "Between Analytic and Empirical," *Philosophy*, XXXII (1957), 112–131; J. L. Mackie, "The Paradox of Confirmation," *B.J.P.S.*, XIII (1963), 256–277. The most complete and lucid essay on the paradox and suggested solutions has been given by Scheffler, *op. cit.*, pp. 236–291.

According to (a), we must count Mb. M'b (i.e., a white swan) as confirming evidence for U '(x)(Mx ⊃ M'x)' (i.e., 'all swans are white'). According to (c), —Mc·M'c (i.e., a non-white non-swan) is irrelevant to U. But U' '(x)(—M'x ⊃ —Mx)' is equivalent to U and since —Mc·—M'c is favorable to U' by (a), according to (d), we must count —Mc·—M'c as favorable evidence for U. Therefore, objects which are neither swans nor white must be counted as favorable evidence for the hypothesis that all swans are white. With respect to (1) and (2), we should say, objects which are neither swans nor white must be counted as favorable evidence for an instance or qualified-instance of the unrestricted universal hypothesis that all swans are white.

That (1) is "hit" by this paradox is apparent. In terms of the discussion above, —Mc·—M'c is apparently irrelevant to the instance of U 'Md ⊃ M'd' but must be counted as favorable because it is favorable to the logically equivalent instance '—M'd ⊃ —Md.' That (2) is also "hit" by this paradox is not quite obvious. The degree of qualified-instance confirmation of h '(x)(Mx ⊃ M'x)' given some evidence e is equal to the degree of confirmation of h' 'M'b' given e and j 'Mb.' H '(x)(—M'x ⊃ —Mx)' is logically equivalent to h. Thus, according to (d), whatever confirms H confirms h as well. To confirm H according to (2) we shall have to have J '—M'b' and determine the degree of confirmation of H' '—Mb' given Je. Presumably then, there will be many apparently irrelevant instances such as —M'b. —Mb which must be counted as favorable for h because they are favorable for the logically equivalent H.

According to Hempel, the paradox is only psychological. Psychologically we are not inclined to consider non-white non-swans as favorable evidence for the hypothesis that all swans are white. However, since the hypothesis asserts that all non-white things are non-swans, there can be no logical reason to suppose a non-white non-swan is not favorable. Thus, from a logical point of view, there simply is no paradox. In *The Logical Foundations of Probability*, Carnap parenthetically notes that he will discuss Hempel's paradox in the now abandoned second volume. "Our results," he writes, "will essentially be in agreement with Hempel's views."[22] In another context, after presenting the paradox roughly as I have presented it, he writes, "Now h and h' are L-equivalent: they express the same law and differ merely in

---

[22] *Ibid.*, p. 496.

their formulations. Therefore, *any observation must either confirm both or neither of them.*"[23] (Italics added.) However, in the paragraph preceding (1) and (2) in Section 2, he says that some individuals in a sample might be "—M(Non-Swan) and hence neither M_1 nor M_2," (i.e., neither disconfirming nor confirming instances). Thus, we cannot be completely confident that Carnap would accept Hempel's view as a method of avoiding Hempel's paradox. So far as I can tell, nothing disastrous would follow for Carnap's view from the rejection of the claim that non-swans are irrelevant in these cases. In fact, Hempel grants that white non-swans must be considered confirming instances and still claims the paradox is only psychological.[24] I suspect then, that the method suggested here for avoiding Hempel's paradox would be acceptable to Carnap.

7. OBJECTION FIVE: THE TENTH 'NECESSARY CONDITION OF ADEQUACY.'

In *Continuum of Inductive Methods*, Carnap introduces the following "necessary condition of adequacy" for any explicatum of probability or degree of confirmation.

C_{10} c is such that for any sentences h and e with any values of k, s, s_2, either $s_2/s \leq c(h, e) \leq 1/k$ or $1/k \leq c(h, e) \leq s_2/s$.[25]

Here s is a sample with s_2 favorable to h and $s - s_2$ unfavorable. k is still the number of Q-predicates in the language, i.e., $k = 2^{pi} = 4$ for the simple language L^2. $1/k$ is the relative width of the single Q-predicate whose occurrence is supposed to be favorable to h.

Now, the following examples show that (1) and (2) violate C_{10}. Consider again the simple hypothesis U '(x)(Mx ⊃ M'x).' The width of $M - M' = w_1 = 1 = w_2 =$ the width of MM'. Let e be a sample with $s_1 = 4$ and $s = 6$. According to (1) we have

$$c_i^*(U, e) = 1 - \frac{4+1}{6+4} = \tfrac{1}{2} = c^*(h, e).$$

But by C_{10} we have

$$\tfrac{1}{4} \leq c(h, e) \leq \tfrac{1}{3}.$$

Thus, (1) violates C_{10}.

---

[23] *Ibid.*, p. 224.
[24] Hempel, "Studies in the Logic of Confirmation," *loc. cit.*, pp. 13–21.
[25] Carnap, *Continuum of Inductive Methods*, p. 26. This condition is still accepted by Carnap. See Carnap, "Replies and Systematic Expositions," *loc. cit.*, p. 993.

To prove (2) violates $CI0$, let ej be a sample with $s_2 = 0$ and $s_1 = 1$. We have

$$c^*_{qi}(M, M', e) = 1 - \frac{1+1}{1+1+0+1} = \tfrac{1}{3} = c^*(h', ej).$$

But by $CI0$ we have
$$0 \leq c(h, e) \leq \tfrac{1}{4}.$$

Thus, (2) violates $CI0$. We have proved then, that (1) and (2) are unacceptable according to Carnap's own adequacy criterion.

Now (2), at least, may be rescued from the above criticism by the method used to avoid Hempel's paradox. If we count, as I have suggested we should, objects that are not M and not M' as favorable for U, then the width of the predicate of individuals that will be favorable for U will be $w_2 = 2$ rather than 1. And, since U asserts that every object is either not M or M', we should count objects that are simply not M as favorable for U – whether or not they are M'. In that case, the width of the predicate of individuals that will be favorable for U will be $w_2 = 3$. For $w_2 > 1$ we must use a generalized version of $CI0$.

(8-1) For any molecular predicate M, any sentences h and e with any values of k, w_2, s and s_2, either $s_2/s \leq c(h, e) \leq w_2/k$ or $w_2/k \leq c(h, e) \leq s_2/s$.[26]

So far as (1) is concerned, nothing has changed because the new width of $w_2 = 3$ does not occur in (1). Thus, our first example proves that (1) violates (8-1). On the other hand, I have not been able to construct a case showing that (2) violates (8-1) so long as we count any object that is either not M or M' as favorable evidence for U. Thus, (2) seems to avoid this fifth objection.

In reply to this fifth objection, someone might claim that the concepts of instance and qualified-instance confirmation are not meant to be explicata of probability$_1$ or degree of confirmation and therefore it is a mistake to apply $CI0$ to (1) and (2). The premise is true but the conclusion does not follow from it. (1) and (2) yield ordinary probability$_1$ or degree of confirmation values for instances and qualified-instances of unrestricted universals. Hence, these values must conform to all of the adequacy conditions introduced for any explicatum of probability$_1$ or degree of confirmation.

Alternatively, someone could claim that $CI0$ should be rejected or,

[26] Carnap, *Continuum of Inductive Methods*, p. 26.

at least, somehow modified. The rejection of C_{10} would mean that a confirmation value might be acceptable even though it is much higher than that warranted by the particular "logical factor" or by the particular "empirical factor" involved. To grant this, we should have to have some justification of the added confidence in the particular hypothesis. On the other hand, there is no reason to reject the claim that C_{10} *might* be adequately modified. It might, for instance, be modified by allowing a certain interval about each of the limits. For example, instead of an upper limit of, say, $\frac{1}{2}$, we might allow $\frac{1}{2}$ plus some small fraction d. Again, however, we should have to have some justification for allowing the value to reach, say, $\frac{1}{2} + d$, if this value was greater than that warranted by the particular logical and empirical "factors" involved.

CHAPTER VI

THE SINGULAR PREDICTIVE INFERENCE

1. INTRODUCTION

In this chapter we shall consider Popper's latest attack on Carnap's system. This attack represents the *fourth* main issue of the dispute between Popper and Carnap. Though Popper writes, "I wish to make it clear that my present criticism, ... though it may incidentally hit Carnap's theory of induction, it is not intended to be read as a criticism of this theory. For I have criticised the probabilistic theory of induction ... in a far more general way in my *Logic of Scientific Discovery*, especially on pp. 390f."[1]

I shall introduce Carnap's theorem for the singular predictive inference in the next section and Popper's objection to it in section three. Popper's objection contains an obvious flaw which is indicated at the end of section three. In section four, a second objection to the theorem is introduced which is inescapable unless one permits the construction of rules or principles to eliminate the use of certain languages. The possibility of constructing such rules has been seriously considered by Keith Lehrer and Wesley Salmon. In section five Lehrer's reply to a type of argument introduced by Salmon to demonstrate the necessity of abandoning all C-functions involving the number of predicates in a language is presented. In section six it is shown that Lehrer's defense of his rule of greater completeness is unsound. Finally, Salmon's objections to the introduction of *any* rules of language selection are presented and criticized. The view defended here is that there are cases in which certain languages should be and are recognized as "privileged" or "preferred" for pragmatic reasons. No single rule is suggested to fit all cases. It is only insisted that

[1] K. R. Popper, "On Carnap's version of Laplace's Rule of Succession," *Mind*, LXXI (1962) 69.

certain languages are particularly useful or appropriate for certain purposes, and that one should not reject *a priori* the possibility of making a reasoned and reasonable selection of one out of a number of languages.

The issue considered in this chapter belongs to the dispute between Popper and Carnap so far as the success or failure of the second objection to the theorem for the singular predictive inference depends on the success or failure of the criticisms offered here of Salmon's view. If the criticisms offered here are satisfactory, the second objection to that theorem may be eliminated. But the issue considered in this chapter has much more significance than the defense of one theorem. For if Salmon's view prevails then every C-function involving the number of predicates in a language should be rejected. Hence, the issue considered here is crucial for the future development of confirmation theories. And so far as the future development of confirmation theories is concerned, the conclusion of this chapter is just this: There is no reason why C-functions should *not* involve or be partially dependent upon the number of predicates in a language. Insofar as this conclusion bears on the *fourth* issue in the dispute between Carnap and Popper, it is favorable for Carnap because it rescues (precariously no doubt) his theorem for the singular predictive inference from my second objection.

2. THE THEOREM FOR THE SINGULAR PREDICTIVE INFERENCE

Carnap calls the inference from one sample to another a "predictive inference" and "the most important special case of the predictive inference is the *singular predictive inference*."[2] Let h be the singular prediction 'Mc.' The relative width of M is w_1/k. e is a sample in which c does not occur and s_1 of s individuals have M. According to Carnap,

$$(1) \qquad c^*(h, e) = \frac{s_1 + w_1}{s + k}$$

In languages with only one primitive predicate, $k = 2^1 = 2$ and (1) becomes indistinguishable from Laplace's Rule of Succession. But (1) is more subtle than Laplace's Rule because (1) does but Laplace's Rule does not vary as the relative width of the predicate in h varies. Carnap notes that Laplace's Rule is more cautious than a "straight

[2] Carnap, *The Logical Foundations of Probability*, p. 568.

rule" method (i.e., a method such that $c(h, e) = s_1/s$). But granting this, neither method takes account of the relative width of the predicate in h. Thus ,(1) is supposed to be preferable to these alternatives.[3]

3. AN ALLEGED CONTRADICTION IN THE SINGULAR PREDICTIVE INFERENCE

Popper tries to show that Carnap's Rule (1) leads to contradictions by introducing the following case. We have a finite set of boxes with n buttons in each. We know each button has the property A or not. We assume that there are as many buttons with A as without A and that there are as many boxes with one A botton, with two A buttons, with three A buttons, etc. Drawing $n - 1$ buttons from one of the boxes, we find they are all A. The question is, what is the probability or degree of confirmation of h 'the next button drawn from this box will be A'? And, according to Popper, we have two incompatible answers. First, since we know the next button will either be A or not, we are interested in the probability of h, given h or not-h. This value must be $\frac{1}{2}$. On the other hand, on the assumption that the relative width of 'A' is $\frac{1}{2}$, (1) yields a value of $\frac{1}{2}$ only if half of our observed sample is A. As Popper remarks, "Thus, for $n = 99$, say, the solution according to Laplace and Carnap is the probability of $h = .99$, rather than $\frac{1}{2}$."[4] Thus, because Carnap would have to grant that the probability of h is, say, .99 *and* $\frac{1}{2}$, Carnap's system must be inconsistent. This is Popper's argument.

Popper's argument has been conclusively refuted by Y. Bar-Hillel[5] and Richard C. Jeffrey.[6] According to Bar-Hillel and Jeffrey, Popper has obtained incompatible values because he has determined the probability of h on different evidential statements. To obtain the value $\frac{1}{2}$, the evidence is 'h or not-h is true.' But to obtain .99, the evidence is '$n - 1 = 98$ buttons are A.' Thus, the apparent inconsistency is easily removed.

---

[3] *Ibid.*
[4] Popper, "On Carnap's Version of Laplace's Rule of Succession," *loc. cit.*, p. 69–72.
[5] Y. Bar-Hillel, "On An Alleged Contradiction in Carnap's Theory of Inductive Logic," *Mind*, LXXIII (1964), 265–267.
[6] Richard C. Jeffrey, "Popper on the Rule of Succession," *Mind*, LXXIII (1964), 129.

4. A SECOND OBJECTION: THE SINGULAR PREDICTIVE INFERENCE VIOLATES CARNAP'S SECOND ADEQUACY CONDITION

As noted in the last chapter, one (the second) of Carnap's conditions of adequacy for any explicatum of probability is

(d)   *L-equivalent hypotheses.* If h and h' are L-equivalent, then C(h, e) = C(h', e).[7]

The following argument proves that (d) is violated by (1), i.e., by the theorem for the singular predictive inference.

Consider a language L^2_{100} with two primitive predicates P and R, and 100 individuals $a_1, a_2, ..., a_{100}$. We have observed e that $s_1 = 1$ of $s = 2$ individuals have PR and we are interested in the degree of confirmation of h, the singular prediction '$Pa_3 \cdot Ra_3$.' That is, we are interested in the degree of confirmation for the hypothesis that some new individual a_3 has both P and R. The relative width of PR is $w_1/k = \frac{1}{4}$. Thus, by (1) we get

(2) $$c^*(h, e) = \frac{1+1}{2+4} = \tfrac{1}{3}.$$

Now consider a language L^3_{100} with the primitive predicates P^1, P^2 and R. We define P^1 and P^2 so their conjunction is equivalent to P. That is,

(3) $$Pa \equiv (P^1a \cdot P^2a).$$

For example, P might be the predicate 'is a bachelor' while P^1 and P^2 are the predicates 'is unmarried' and 'is male', respectively. Again we observe e that $s_1 = 1$ of $s = 2$ individuals have P^1P^2R and we are interested in the degree of confirmation of h', the singular prediction '$P^1a_3 \cdot P^2a_3 \cdot Ra_3$.' The relative width of P^1P^2R is $w_1/k = \frac{1}{8}$. Thus, by (1) we get

(4) $$c^*(h', e) = \frac{1+1}{2+8} = \tfrac{1}{5}.$$

Then, since h ≡ h' according to (3), and $\tfrac{1}{5} \neq \tfrac{1}{3}$, (1) clearly violates (d). Thus, (1), the theorem for the singular predictive inference, must be rejected because it violates Carnap's second condition of adequacy (d).

---

[7] See Chapter V, Section 6.

This *kind* of objection might be eliminated if one had some principle by means of which one could reject certain languages in favor of others. If we had, for example, a rule prescribing *only* the use of, say, L^3_{100} in cases like the one above, we could prevent the violation of (d). The serious consideration of such rules involves not only the development of the dispute between Carnap and Popper but, as the next section shows, the development of future confirmation theories.

5. THE PROBLEM OF LINGUISTIC VARIANCE

In "Descriptive Completeness and Inductive Methods," Keith Lehrer argues that Wesley Salmon's attempt to show that a certain class of C-functions yield incompatible results is unsuccessful.[8] The C-functions involved are characterized by their being functions of the number of Q-predicates in the language in which the C-function is used. Lehrer uses the following case to illustrate the problem.

Consider a language L^{RWB} which contains three predicates, R, W, and B, that are logically exclusive in paris and exhaustive, and a language L^R which contains only R. We have a fair die with two red sides, two white sides and two blue. Let h_R be the hypothesis that on the n*th* toss of the die a red face will turn up. Letting t stand for any tautology, we have

(a) $\qquad c(h_R, t)$ in $L^{RWB} = \frac{1}{3}$
(b) $\qquad c(h_R, t)$ in $L^R \quad = \frac{1}{2}$.

These results are incompatible.[9]

Now, Salmon has argued that any C-function or inductive rule must be judged by the following *criterion of linguistic invariance* (CLI).

Whenever two inductive inferences are made according to the same rule, if the premises of the one differ purely linguistically from the premises of the other, then the conclusion of the one must not contradict the conclusion of the other.[10]

---

[8] *The Journal of Symbolic Logic*, XXVIII (1963), 157–160. Lehrer considers Wesley Salmon's argument in "Vindication of Induction," *Current Issues in the Philosophy of Science*, ed. H. Feigl and G. Maxwell (New York: Holt, Rinehart and Winston, 1961), pp. 245–256. But similar arguments may be found in Salmon's "On Vindicating Induction," *Philosophy of Science*, XXX (1963), p. 254, and "Inductive Inference," *Philosophy of Science: The Delaware Seminar*, ed. B. Baumrin (New York: Wiley, 1963), Vol. II.

[9] *Ibid.*, pp. 157–158.

[10] Salmon, "On Vindicating Induction," *loc. cit.*, p. 254. CLI has roughly the same effect as Carnap's first adequacy condition in *The Logical Foundations of Probability*, p. 285, and Hempel's "Equivalence Condition" in "The Logic of Confirmation," *loc. cit.*, p. 110.

THE SINGULAR PREDICTIVE INFERENCE 65

Thus, the type of C-function illustrated above must be rejected, according to Salmon, because it violates CLI.

Lehrer evidently accepts CLI (since he is obviously unhappy with the incompatible results above) but believes Salmon's solution is too costly. According to Lehrer, the following *rule of greater completeness* (RGC) should be used instead.

For any two languages L_j and L_k having the same individual constants such that two sentences e and h occur in both languages and c(h, e) in $L_j \neq$ c(h, e) in L_k, select the language L_{jk} to calculate c(h, e) having the same individual constants as L_j and L_k.[11]

The crucial (and dubious) point of Lehrer's argument for the adoption of RGC concerns his claim that, in the illustrative case above, the result given in (a) is intuitively correct while the result given in (b) is counterintuitive.[12] If Lehrer's intuition is accurate then the following further claim might be justified. "Since it is clear," he writes "that the greater descriptive completeness of L^{RWB} as compared to L^R ... accounts for the difference in the results, the only conclusion that would seem justified is that the use of ... C-functions of the kind in question should be restricted by some rule to languages of greater rather than lesser descriptive completeness."[13]

6. REFUTATION OF LEHRER'S ARGUMENT

Though I suspect Lehrer's intuition has led him astray, I shall not challenge that intuition directly. Instead, I shall reconstruct Lehrer's case with a slight alteration to show that it is possible to have a *less* complete language yield a value that is intuitively more acceptable than the value yielded by a *more* complete language. Consider a case in which all things are as in Lehrer's case except that the die has three red sides, two blue sides and one white. Again we have

(a) $\quad\quad\quad c(h_R, t)$ in $L^{RWB} = \frac{1}{3}$

(b) $\quad\quad\quad c(h_R, t)$ in $L^R \quad\; = \frac{1}{2}$.

But in this case Lehrer would have to say that (a) is counterintuitive and (b) is intuitively correct even though L^{RWB} is *more* complete than L^R. Lehrer would have to say (a) is counterintuitive in this case

---

[11] Lehrer, "Descriptive Completeness and Inductive Methods," *loc. cit.*, p. 158.
[12] *Ibid.*
[13] *Ibid.*

for the same reason he claims (b) is counterintuitive in his case, namely, that we *know* R has only two chances in six of turning up in this case. Thus, if Lehrer's intuitive sense is consistent, we have a case which is such that the application of RGC to it would yield an unacceptable selection, viz., L^{RWB} rather than L^R.

Now that we have shown that Lehrer's defense of RGC is not acceptable, consider briefly the fundamental error committed by Lehrer. The error is simply this: Lehrer compares the confirmation values obtained using tautological evidence with those obtained using both tautological *and* nontautological evidence, and then rejects one result obtained using tautological evidence because it is not near enough to a result obtained using both tautological *and* nontautological evidence. I shall explain this more fully. In Lehrer's case, the degree of confirmation of h_R given t (in L^R) is $\frac{1}{2}$. The degree of confirmation of h_R given t *and* the information about the distribution of the colors on the die (in L^R) is (or we suppose *should* be)nearer to $\frac{1}{3}$. Lehrer's intuitive judgment, that is, is simply a judgment based on evidence that is different from the evidence on which his *non*intuitive judgment is based. Without introducing such additional evidence Lehrer would have no basis for his suspicion. But once it is pointed out that his suspicion arises only because he has introduced additional evidence for his intuition to consider, Lehrer's argument collapses.

Now, while the objection which has just been raised does show that Lehrer's defense of RGC is unsound, it does *not* show that RGC is unacceptable. For in order to turn the tables on Lehrer, an argument that is formally identical to his was used. Consequently, his mistake was made again, i.e., again an appeal was made to "illicit additional information" about the distribution of colors on the die. Thus, for all we know RGC may indeed be acceptable, though so far we are unable to prove it.

7. METAPHYSICAL VERSUS PRAGMATIC REASONS FOR CON-CONSTRUCTING RULES TO ELIMINATE CERTAIN LANGUAGES

Although, like Lehrer, I can see some point in trying to develop plausible rules or principles for the selection of appropriate languages in which to conduct one's particular research, Salmon rejects *every* rule of this sort as well as *every* C-function involving the number of predicates in a language. He prefers his rather drastic solution to the problem for two reasons. He believes, in the first place, that rules such as RGC have the effect of forcing us "to regard one particular language with its particular set of predicates as privileged" and this, he claims, constitutes "an outrageous bit of metaphysics."[14] In the second place, he believes that C-functions which include some essential reference to the number of predicates in a language have the effect of allowing "arbitrary features of the choice of language to influence predictions concerning empirical fact."[15] To put it another way, he writes,, "if the task to be accomplished is the prediction of objective fact, then we do not want to adopt a rule which reflects the arbitrary features of the choice of language in its results."[16] I shall reply to each of these claims.

Consider the first claim, namely, that such C-functions involve us in "an outrageous bit of metaphysics." The examples introduced above *seem* to support part of this claim. The judgment regarding the acceptability of one of the two languages clearly involved some reference to and assumption about the type of world to which the languages were applied. If whenever one makes such assumptions one is guilty of "playing outrageous metaphysics," then we must grant Salmon's view. But there are two good reasons for questioning this view. First, in artificial worlds or games of chance we very often *know* exactly how the world to which our language is applied is designed. We know, for example, that the die has so many white faces, so many red and so many blue; or, we know an ordinary die has six different faces; or, we know how the faces in a deck of cards are colored; or, we know an ordinary penny has two different faces. In such cases there seems to be nothing outrageous about our assumptions or references to the world. Indeed, it would be more outrageous to simply ignore some of our knowledge, i.e., to ignore some of our evidence concerning the outcomes of any predictions involving these worlds. Second, aside from artificial worlds or games of chance, it is by no means obvious that whenever one makes an assumption about the world to which one's language is applied, one is "playing outrageous metaphysics." Such assumptions *could* be made quite tentatively (i.e., instead of dogmatically) and for pragmatic reasons. Such assumptions would be synthetic *a priori* (hence metaphysical) suppositions made for certain immediate purposes, but not for all time (hence, not dogmatically). For example, suppose we are interested in making

---

[14] Salmon, "On Vindicating Induction," *loc. cit.*, p. 260.
[15] *Ibid.*, p. 249.
[16] *Ibid.*, p. 256.

predictions about the colors of certain objects in some region. We might assume a language containing predicates designating only the primary colors because no other distinctions are relevant. Other distinctions might be irrelevant for a number of reasons. We might be interested in the likelihood of selling paint of a certain color in that region; or, we might be interested in discovering how often (or whether or not) the physical conditions of the area permit the observation of the primary colors (i.e., perhaps the region is covered with smog most of the time); or, we might be interested in discovering how many people living in the region can see just these colors (i.e., assuming now that the physical conditions of the region are fairly normal). These examples should be sufficient to illustrate our point: It is quite possible to make assumptions about the world that are *tentative* and *useful*, and not outrageous *or* "outrageously metaphysical."

Now consider Salmon's second claim, namely, that such C-functions allow "arbitrary features of the choice of language" to "influence predictions concerning empirical fact." I have already suggested three cases in which there seem to be very good pragmatic reasons for preferring some languages to others. Thus, there certainly are cases in which the relevant features of the language adopted are not just "arbitrarily" selected. But the extremely dubious part of Salmon's claim is his implicit assumption concerning the neatness of the distinction between language and empirical facts.[17] It is trivially true that without a language we can not say what the facts are. And it is not quite as trivial to point out that, aside from artificial worlds and languages, it is at least difficult to know exactly what part or how much of what we call the "empirical facts" are *just that* because of the language we happen to use. At any rate, my intention here is only to call the reader's attention to more thorough investigations of this problem.[18] As long as the distinction between language and empirical fact is not a neat one, we can not place much weight on Salmon's second claim. Therefore, unless the replies offered to Salmon's objections to constructing rules to eliminate the use of certain languages for certain purposes are faulty, there is no reason to rule out the possibility of constructing such rules. That is, first, there is no reason to

---

[17] See Salmon, "Vindication of Induction," *loc. cit.*, p. 249 and p. 256.
[18] For discussion of this extremely problematic issue, see N. R. Hanson, *Patterns of Discovery* (Cambridge: University Press, 1958), Chapter I; P. K. Feyerabend, "How to be a Good Empiricist – A Plea for Tolerance in Matters Epistemological," in Baumrin, *op. cit.*, and Popper, *Conjectures and Refutations*, pp. 42–48.

suspect that the second objection to the theorem for the singular predictive inference can not be eliminated. And, more importantly, there is no reason to suspect that every C-function involving the number of predicates in a language must lead to incompatible results.

CHAPTER VII

LAKATOS ON APPRAISAL, GROWTH AND ANALYTIC GUIDES

1. INTRODUCTION

In the second volume of the Proceedings of the International Colloquium in the Philosophy of Science at London in 1965, Imre Lakatos has an extremely interesting historical critique of the dispute we have been analyzing since Chapter II.[1] Anyone reading his essay and this book will notice almost immediately that there are significant differences between our two accounts of the main problems and their solutions. In this chapter I shall consider three main points of contention and a few ancillary issues. First it is shown that there is a difference in the aims of logicians like Carnap and methodologists like Popper, and that Lakatos's view of the nature of the "appraisal" problem presumably attacked by Carnap and Popper is hopelessly ambiguous. In section three I begin by disclosing an ambiguity in Lakatos's view of the nature of the "growth" problem and then show that, contrary to Lakatos's claims, a Carnapian logic of discovery is *not* "atheoretical" or "acritical". Finally, it is shown that analytic sentences may very well function as guides of life, as both Carnap and Popper have asserted.

2. 'THE' PROBLEM OF GRADING THEORIES

According to Lakatos, the breakdown of "classical empiricism" left would-be empiricists with two problems, namely, (1) *"the appraisal of conjectural knowledge"* and (2) *"the growth of conjectural knowledge."*[2] The result was the development of two empiricist schools, that of *"neoclassical empiricism"* (Johnson, Broad, Keynes, Ramsey, Jeffreys,

---

[1] Lakatos, "Changes in the problem of inductive logic," *op. cit.*
[2] *Ibid.*, p. 322.

LAKATOS ON APPRAISAL, GROWTH, ANALYTIC GUIDES 71

Reichenbach and Carnap) and that of *"critical empiricism"* (Popper and later popperites presumably, though Lakatos never provides us with a list). The trouble with the "two problems" cited is that each seems to designate a different problem for neoclassicists (especially Carnap) and Lakatos, and occasionally for Lakatos himself at different times. I will comment on problem (1) in this section and (2) in the next, with some overlapping.

With respect to (1), *"the appraisal of conjectural knowledge,"* Lakatos claims that "the original problem of confirmation was no doubt the confirmation of *theories* rather than that of *particular predictions*" and that "it was agreed that a crucial adequacy requirement for confirmation theory was that it should *grade* theories according to their evidential support."[3] The first of these claims seems very dubious if it is supposed to apply to each of the neoclassicists listed above and the second is hopelessly ambiguous. The first claim, for example, seems to fit Harold Jeffrey's position fairly well, for Jeffreys was inclined to use the terms 'induction' and 'generalization' as synonyms.[4] But I can find no justification for the first claim in Keynes, Broad or Reichenbach, and it is certainly not the driving force behind Carnap's *magnum opus* on probability. As Carnap explains it, "The major aim of this book ... is the actual construction of a system of inductive logic ...."[5] He believed, wrongly no doubt, that "all inductive reasoning, in the wide sense of nondeductive or nondemonstrative reasoning, is reasoning in terms of probability" and that "hence inductive logic, the theory of the principles of inductive reasoning, is the same as probability logic."[6] These remarks are very much in the spirit of Keynes and the other neoclassicists. They were interested in providing what many of us still hope to see, namely, a thoroughly developed system of inductive logic similar in rigor to that of deductive logic. An account of inductive generalization, whether the generalization is lawlike, theorylike, or accidental would be an important part of the system, but hardly that for the sake of which the whole "research program" exists. Thus, there is a fundamental difference of *intent* here between the neoclassicist logicians and the critical empiricist methodologists, a difference that Lakatos never appreciates. Lakatos reads a sentence

---

[3] *Ibid.*, p. 330.
[4] Harold Jeffreys, *Theory of Probability* (Oxford: The Clarendon Press, 1939), p. 1.
[5] Carnap, *The Logical Foundations of Probability*, p. v.
[6] *Ibid.*

like this:

> "When scientists speak about a scientific *law or a theory, or also a singular statement*, for example, a prediction, on the one hand, and certain observational data or experimental results, on the other, they often state a relation between those items in forms like these: a. 'this experiment again confirms the theory T' (or: '... supplies new evidence for ...')."[7] (Italics added.)

and interprets it thus:

> "... his theory of probability was to solve the time-honoured problem of induction, which *according to Carnap, was to judge laws and theories* on the basis of evidence."[8] (Italics added.)

The logician speaks of laws, theories and singular statements, in effect *all* kinds of statements, but the methodologist only hears a claim about laws and theories. Consequently, the methodologist, Lakatos, mistakenly identifies the "original problem of confirmation" for neoclassicists like Carnap.

The ambiguity of Lakatos's account of "the" problem of grading "theories according to their evidential support" is at least fourfold. (i) Sometimes he regards it as a problem that logically could be solved by constructing a measure function for the assignment of numerical values greater than zero to theories, *whatever the value measures*. "Why did he [Carnap] not at least *try*," he writes "perhaps following the Wrinch-Jeffreys idea of a simplicity ordering (expounded in 1921), to introduce *immediately* a system with positive measures for universal propositions."[9] (ii) At other times he seems to regard it as a problem that logically could be solved only by constructing a measure function for the assignment of numerical *probability* values greater than zero to theories. The "radical problem-shift" he refers to in the sentence immediately preceeding the one just quoted[10] is apparently a shift away from the problem as thus defined. (iii) At still other times he seems to regard it as a problem that logically could be solved by showing how some kind of probability values could be used to grade theories. For example, he tells us that "a theory of confirmation assigns marks – directly or indirectly – to theories, it gives a *value-judgement*, an *appraisal* of theories" and in a footnote explains the

---

[7] *Ibid.*, p. 1.
[8] Lakatos, *loc. cit.*, p. 350.
[9] *Ibid.*, p. 339.
[10] *Ibid.*

term 'indirectly' with the remark "indirectly with the help of qualified instance confirmation."[11] (iv) Finally, he occasionally writes as if he believed that the problem logically could be solved merely by finding some plausible procedure for determining which theories are in some significant sense better. "The excess [content of rival theories] is in no sense measurable" he insists, but it is crucial for determining which of two theories is better.[12]

In the presence of such ambiguity, most of what Lakatos has to say about changing "research programmes," "problem-shifts" and "problem-solving" has little value. If, for example, "the" problem of grading theories is taken to be that explained in (ii) then, contrary to Lakatos's claim, the school of "critical empiricism" could not show that their solution to the second problem of empiricism, *"the growth of conjectural knowledge,"* "solves the most important aspects of the first too."[13] They could not show this because they, or at least Popper himself, have insisted that the probability of an unrestricted universal is zero "for *all* 'acceptable' probability functions."[14] On the other hand, if "the" problem of grading theories is taken to be that explained in (iii) then, contrary to Lakatos again,[15] Carnap's use of the theorems of instance and qualified-instance confirmation does not constitute a problem shift. It would have to count as a *solution*, an unsatisfactory one to be sure, but nevertheless a solution. Thus, either or both of the above two claims of Lakatos must be abandoned: Either Carnap made no "radical problem-shift" or the "solution" Popper and Lakatos have provided to the problem of "growth" is irrelevant to Carnap's problem of "appraisal." In my opinion both claims are false, because for Carnap the problem of "appraisal" has always been that explained in (iii). Thus, there was no "problem-shift" and Popper's "solution" is irrelevant to Carnap's problem.

3. THE PROBLEM OF THE GROWTH OF KNOWLEDGE

With respect to (2), *"the growth of conjectural knowledge,"* Lakatos seems to hold the view that *"Carnap's neoclassical solution* to the "appraisal" problem ... *leaves the problem of discovery, the problem of the growth of knowledge, untouched."*[16] But then he also seems to think

[11] *Ibid.*, p. 343.
[12] *Ibid.*, p. 379.
[13] *Ibid.*, p. 322.
[14] *Ibid.*, p. 332.
[15] *Ibid.*, pp. 338–339.
[16] *Ibid.*, p. 326.

that this is impossible, for replying to Bar-Hillel's suggestion of a "division of labour, the Carnapians concentrating mostly on a rational *'synchronic'* reconstruction of science and the Popperians remaining mostly interested in the *'diachronic'* growth of science,"[17] Lakatos insists that *"this division of labour seems to imply that the two problems are somehow independent. But they are not. I think the lack of recognition of this interdependence is an important short-coming of logical empiricism in general and of Carnap's confirmation theory in particular."*[18] Again, however, one must insist that Lakatos cannot have it both ways: *Either* the problems are independent and it is logically possible for Carnap to contribute to a solution to one and not the other, *or* the problems are not independent and it is not logically possible for Carnap to contribute to a solution to one and not the other. I think that Lakatos's "real" view is that Carnap (and neoclassicists generally) has at best a perverse solution to the "growth" problem, so perverse that it does not merit the quasi-respectability it would receive if it were called a solution. But whatever position one takes with respect to Lakatos's psychology, the untenability of his logic is unquestionable.

The reason I think that Lakatos's "real" view is that Carnap has a perverse "logic of discovery" is that Lakatos has a critique of *"A Carnapian logic of discovery."*[19] In that critique Carnap's "heuristic" is described as a process generally known as 'Bayesian conditionalization' and consisting of the following three steps. (a) An appropriate language is chosen for the expression of all hypotheses. (b) A hypothesis is chosen. (c) Its degree of confirmation relative to the total available evidence is determined. Although these three steps are fundamental for 'Bayesian conditionalization,' this "heuristic" falls seriously short of what I would regard as a complete Carnapian logic of discovery. The latter would require some reference to the determination of the estimated or expected utility of any hypothesis under consideration. However, because Lakatos's critique ignores the problems involved in assigning utility values to hypotheses, so will mine. The latter problems are discussed in detail in Chapter IX.

The trouble with Bayesian conditionalization, according to Lakatos, is that it is supposed to be *"atheoretical"* and *"acritical."*[20] Further-

[17] *Ibid.*, p. 329.
[18] *Ibid.*, pp. 329–330.
[19] *Ibid.*, pp. 346–349.
[20] *Ibid.*, p. 347.

more, without theories there cannot be any explanation, and without criticism there can be no "potential falsifiers" and, consequently, no testability![21] In short, as far as Lakatos is concerned the adoption of "a Carnapian logic of discovery" would be nothing short of catastrophic for scientific practice. Unfortunately (for Lakatos but not for Carnapians) Lakatos is mistaken about the "atheoretical" and "acritical" nature of Bayesian conditionalization and, therefore, also about the additional inferences he draws with respect to explanation and testability. I shall defend each part of this assertion in turn.

Why does Lakatos believe that Carnap "had to abandon *any* reference in his theory to universal propositions?"[22] There seem to be two reasons. In the first place, Lakatos seems to believe that Carnap's theorems of instance and qualified instance confirmation have nothing to do with "confirmation."[23] This is plainly false because the two theorems are defined in terms of Carnap's explicatum degree of confirmation. As I have already shown, those theorems are unacceptable. Nevertheless, they are part of the system and their function is to "grade" generalizations on the basis of the degree of confirmation of their "instances" and "qualified instances."

Secondly, Lakatos seems to believe that Carnap "had to abandon *any* reference in his theory to universal propositions" because he wanted to appeal to the Ramsey-Definetti theorem and "the proof of this theorem hinges on the lemma that $p(h) \neq 0$ for all contingent propositions."[24] The premisses of this argument are true, but they do not imply the conclusion that Lakatos thinks they imply. Whether or not Carnap wants to appeal to the Ramsey-DeFinetti Theorem (and I grant that he does), his explicatum degree of confirmation is such that in universes with an infinite number of individuals, "universal propositions" always have a confirmation value of zero. Thus any *direct* appraisal of such propositions is, as Lakatos insists and no one doubts, "trivial" and useless for "grading" purposes. Hence, Carnap does not appraise such propositions *directly*. Indeed, the *direct appraisal* of such propositions is "excluded" and "abandoned," to use Lakatos's favorite terms. But there is clearly a big difference between the abandonment of the direct appraisal of generalizations and the abandonment of the generalizations themselves. The premises of

---

[21] *Ibid.*, p. 348.
[22] *Ibid.*, p. 336.
[23] *Ibid.*, p. 335.
[24] *Ibid.*, pp. 336–337.

Lakatos's second argument merely lead to the exclusion of a kind of appraisal procedure, not to the exclusion of a kind of object of appraisal.

Although Lakatos does not seem to have any reply to my objection to his first argument, he does have a reply to my objection to his second argument. It is just this: "... in calculating rational betting quotients of particular hypotheses one cannot escape appraising (genuine universal) *theories*."[25] He obtains this conclusion from roughly the following two premisses. (a) Carnap cannot calculate the rational betting quotient (= logical probability) of any "particular hypothesis" without choosing a "correct language," and (b) he cannot choose a "correct language" without, at the same time, choosing "the language of the most advanced, best-corroborated theory."[26] The first premiss is, of course, true, but the second is demonstrably false. As I have suggested in Chapter VI, Section seven, the selection of a language is and ought to be determined largely by one's purposes. For "pure scientists" we may assume that if all other things are equal, their "correct language" is one which permits the most accurate description of the "real world," the world they hope to be learning more and more about as science improves. But it is logically possible that "the language of the most advanced, best-corroborated theory" is *not* the one which permits the most accurate description of the "real world." As Lakatos explains it:

> "The successive scientific theories may be such that each increase of truth-content could be coupled with an even bigger increase in hidden falsitycontent, so that the growth of science would be characterized by increasing corroboration and decreasing verisimilitude. Let us imagine that we hit upon a true theory T_1 (or on one with a very high verisimilitude); in spite of this we manage to 'refute' it with the help of a corroborated falsifying hypothesis f, replace it by a bold new theory T_2 which again gets corroborated, etc., etc. Here we would be ... moving even further from the truth while assuming that we are soaring victoriously toward it. Each theory in such a chain has a higher corroboration and a lower verisimilitude than its successor: such is the result of having 'killed' a true theory."[27]

Clearly then, Lakatos recognizes an important difference between choosing a "correct language" and choosing "the language of the most advanced, best corroborated theory." But even if he would reject his claims in the long quotation above, the logical distinction

[25] *Ibid.*, p. 360.
[26] *Ibid.*, pp. 362–364.
[27] *Ibid.*, p. 397.

required to falisfy the second premiss of the argument we are considering is there for all to see. I conclude, therefore, that Lakatos's (presumed) reply to my objection to his second argument for his view that Bayesian conditionalization is "atheoretical" is unsound. Furthermore, since that "heuristic" has now been shown to be perfectly compatible with rather than exclusive of laws and theories, it follows that it does not militate against the use of such generalizations at all. Hence, half of Lakatos's claim about Carnap's logic of discovery is false.

Consider now, the other half of Lakatos's claim, namely, that Carnap's logic of discovery is "acritical" and eliminates "testability." Why does Lakatos believe Bayesian conditionalization is "acritical?" Again there seem to be two reasons. First, he recognizes as many others have *rightly* recognized, that the Bayesian approach is in an important but not entirely clear sense, *closed*. "The learning process" he insists "is strictly confined to the original prison of the language,"[28] One begins with a set of mutually exclusive and exhaustive hypotheses and assigns probability (or probability *and* utility) values in much the same way that rewards are distributed in a zero-sum game. In the latter, what one player wins, another player loses. Here, an increase in the probability (or expected utility) of one hypothesis is bought at the expense of a decrease in the probability (or expected utility) of another, since there is exactly one unit of probability to be distributed among all contenders. Thus, the only kind of "criticism" that seems to be possible is that of raising or lowering the *value* of the share allotted to each hypothesis of the given set. But, the critics rightly insist, scientific inquiry often reveals a kind of criticism that is much more thoroughgoing, much more severe. Instead of having mere changes in the probability (or expected utility) values allotted to each member of a given set of competing hypotheses, one finds, in Popper's favorite phrase, "bold new theories," even "fact correcting" theories.[29] Instead of an "open society" of hypotheses taking turns as it were becoming "king for a day" (or a scientific era), one finds something more like a war between foreign adversaries. The latter kind of "critical ethos" is described very forcefully by Lakatos. "Evidential support for a theory" he says, "obviously depends not just on the *number* of the corpses of its rivals. It also depends on the *strength* of the killed. That is, evidential support is, as it were, a hereditary

[28] *Ibid.*, p. 347.
[29] *Ibid.*, p. 320.

concept: it depends on the total number of rival theories killed by the killed rivals."[30] Although I suspect the "evidential support" referred to in these remarks depends on even more than Lakatos seems to be allowing, the only point I wish to emphasize about the view he is expressing is the "depth" or "severity" of the kind of criticism he is suggesting. It is clearly intended to go beyond anything typically covered by the model of Bayesian conditionalization, and it is absolutely necessary if one is going to properly conceptualize scientific inquiry. But, contrary to the implication of Lakatos's claim, the "warlike" criticism is *not* the only kind. There is also a place for a milder kind of criticism, a kind of criticism that adherents of Bayesian conditionalization have attempted to "rationally reconstruct" and that philosophers like Thomas Kuhn have cited as characteristic of "normal science." The following exerpt from Kuhn, for example, suggests a clear appreciation of the existence and role of a kind of criticism that Lakatos, like Popper, fails to perceive or conceptualize.

> "Few people" Kuhn writes, "who are not actually practitioners of a mature science realize how much mop-up work ... a paradigm leaves to be done or quite how fascinating such work can prove in the execution.... Mopping-up operations are what engage most scientists throughout their careers. They constitute what I am here calling normal science.... that enterprise seems an attempt to force nature into the preformed and relatively inflexible box that the paradigm supplies. No part of the aim of normal science is to call forth new sorts of phenomena; indeed those that will not fit the box are often not seen at all. Nor do scientists normally aim to invent new theories, ... Instead, normal scientific research is directed to the articulation of those phenomena and theories that the paradigm already supplies.
> Perhaps these are defects .... The enterprise now under discussion has drastically restricted vision. But those restrictions... turn out to be essential to the development of science.... And normal science possesses a built-in mechanism that ensures the relaxation of the restrictions that bound research whenever the paradigm from which they derive ceases to function effectively."[31]

If Kuhn is right about the existence and role of "normal science" in research, and if I am right in thinking that the kind of criticism that occurs in "normal science" is precisely the sort that adherents of Bayesian conditionalization are attempting to "rationally reconstruct," then Lakatos's claims that Carnap's logic of discovery is "acritical" and "eliminates testability" are false. It seems to me that both premises of this argument are by now beyond reasonable doubt.

---

[30] *Ibid.*, p. 395.
[31] Thomas S. Kuhn, *The Structure of Scientific Revolutions* (Chicago: University of Chicago Press, 1962), p. 24.

Hence, the other half of Lakatos's claim about Carnap's logic of discovery is flase.

Finally then, insofar as Lakatos's "real" view about Carnap's solution to problem (2), *"the growth of conjectural knowledge,"* rests on the unsound arguments criticized in this section, that view is entirely unwarranted. Thus, the upshot of my analysis of the last two sections is just this: Lakatos's account of the neoclassicists' and especially Carnap's views of the two problems of "appraisal" and "growth" are for the most part seriously muddled, misleading and mistaken. Let us then turn to another issue.

4. ANALYTIC AND OTHER 'DUMB' GUIDES OF LIFE

In a somewhat famous paper A. J. Ayer claimed that "in the sense in which probability is the guide of life ... [I do not] ... think that statements of probability can be logically true."[32] Even if such a statement were based on all available evidence, he claimed that "there is no sense ... in which the proposition ... can be superior to any of the others as a measure of probability. And this being so, there can be no practical reason why we should take it as a guide."[33] In the discussion following his paper Professor Ayer was supported by P. K. Feyerabend.[34] Since then others have joined the chorus. In particular, Wesley C. Salmon has written that he has been "unable to see, in spite of Carnap's patient efforts to explain, how such analytic statements could constitute 'a guide of life.'"[35] Similarly, W. C. Kneale wrote "if confirmation theory is to be the guide of life, confirmation itself must be defined by reference to the rationality of action in the future. But I do not see how in that case statements of the form $c(h, e) = p$ can ever be analytic."[36] Finally, Lakatos adds his support to these claims, saying "if Popper's appraisal of scientific theories is analytic, then Popper cannot explain how science can be a guide of life."[37]

Contrary to these authors, and in support of *both* Carnap and

---

[32] A. J. Ayer, "The conception of probability as a logical relation," *Observation and Interpretation in the Philosophy of Physics*, ed. S. Korner (New York: Dover Publications, Inc., 1957), p. 17.
[33] *Ibid.*, p. 14.
[34] *Ibid.*, p. 21.
[35] W. C. Salmon, "The justification of inductive rules of inference," *The Problem of Inductive Logic*, ed. I. Lakatos (Amsterdam: North Holland Pub. Co., 1968), p. 40.
[36] *Ibid.*, "Confirmation and Rationality," p. 61.
[37] Lakatos, *loc. cit.*, pp. 401–402.

Popper, I shall argue that analytic statements *do* as a matter of fact function as guides and that *a fortiori* they certainly *can* have such a function. Moreover, I shall show that some analytic statements are better guides than others.

It is easy to see why someone would want to claim that analytic statements cannot be *prescriptive*. After all, they are (I would grant) vacuous, uninformative or completely void of empirical content. So if they do not tell us anything, they cannot tell us to do anything either. They cannot provide us with information as to how we should act because they cannot provide us with any information at all.

I think that this argument is absolutely sound. However, it is beside the point because guides of life are not *identical* to prescriptions. Or, to put this fundamental point in other words, many things function as guides without prescribing anything. Consider the following examples.

Statements (1) and (2) are analytic in the most straightforward sense of this difficult term, i.e. they have self-contradictory denials.

(1) 'p and q' truth-functionally implies 'p'.
(2) 'p' truth-functionally implies 'p or q'.

Nevertheless each of these statements has been and shall continute to be used by students of logic as a guide in the derivation of conclusions from premisses. (1), for example, may be used to guide a student to 'A' from 'A and B.' The student examines the patterns of (1) and of his premiss 'A and B,' and writes down 'A' as his conclusion. Because (1) is analytic, it cannot prescribe anything. But the for very same reason, it may be used as a completely reliable guide for the student's behavior. Given 'A and B' and some instructions (including prescriptions probably) about how to use statements like (1) in demonstrations, the student knows what to do, i.e., how to act or behave with respect to 'A'. Similarly, the student would be justifiably confident that 'A or B' is one of the statements that may be validly derived from 'A' given (2). Sentence (2) is his guide to the conclusion 'A or B' from the premise 'A' just as surely as if (2) were an honest man giving him directions verbally. But unlike the honest man, (2) is mute. It literally cannot speak; it cannot tell anyone anything.

There are many other kinds of "dumb" guides. The lines on the paper on which I am now writing have no cognitive content. Hence they are totally uninformative in any cognitive sense. But they are being relied upon as guides to keeping my written sentences more or less parallel to one another. The sights on rifles function as guides to accurate shooting. Road maps would be not only useless but self-defeating if regarded as some sort of prescriptive device. They would apparently be prescribing us to go every which way from this town to that. On the other hand, they are excellent guides. Mathematical tables of square roots, common logarithms, trigonometric ratios, etc. do not prescribe anything, but are useful guides for behavior. We use them in order to find out what numerical expression should be plugged into this or that formula.

These examples should be sufficient to demonstrate the truth of my claim that many things function as guides without prescribing anything. Analytic statements are merely one of the kinds of things that can play the role of "dumb" guides.

Interesting as this conclusion is, it is not the only point worth making about such guides. As I mentioned in the second paragraph, some analytic statements are unquestionably better than or preferable to others. For example, suppose a student is given (1) and (2), the premiss 'A' and is asked to derive 'A or B.' (1) and (2) are both analytic, but (1) is worthless for the student's problem. (2) is better than or preferable to (1) because (2) may be used as guide to the desired conclusion, while (1) cannot be used given the premise 'A' and the instruction to derive 'A or B.' This is just a trivial case of a fairly common type of occurrence among students of logic. Given a whole battery of useful rules, axioms, theorems and definitions, usually only some of them are useful for a given job. Recall the frequency with which this or that complicated substitution is made in a given theorem only to find that not only has the desired theorem failed to appear but that what has appeared is farther from the desired result than some line before it. While we are carrying out the complicated substitution we are fully conscious that we must follow a certain path, we must repeat this symbol here and that one there. Exactly what this path is depends on what theorem, axiom, definition or rule we happen to be applying.

Similar remarks could be made with respect to maps, mathematical tables, various charts and graphs, etc. A completely accurate table of square roots is completely worthless as a guide to trigonometric ratios. The periodic table cannot guide our inferences in the calculus of sentences. All guides have limited applicability, and many of them have equally limited reliability. The so-called synthetic probabilistic inference rules are in the latter class. They are not only liable to yield a conclusion that is true but irrelevant, but they may well yield a false conclusion from true premisses.

It might be objected that the distinction between guides and prescriptions that I have been insisting upon here is *ad hoc* and misleading. It is *ad hoc* because it is merely the result of trying to save logically true probability statements. It is misleading because the two words are usually regarded as interchangeable; so I am drawing a line where it has not existed previously. The first objection is simply false and would have been *ad hominem* at best anyhow. The second is, I think, very near the truth. The two words are usually used interchangeably. But, as I have tried to illustrate with numerous examples, they do not have *identical* intensions or extensions. Furthermore, the differences seem to be so great that once they are brought to one's attention, it is difficult to imagine any advantage in allowing them to remain suppressed.

CHAPTER VIII

HINTIKKA AND HILPINEN ON INDUCTIVE GENERALIZATION

1. INTRODUCTION

In this chapter I shall examine the "heir apparent" of Carnap's system of inductive logic, namely, that constructed by Jaakko Hintikka and developed further with Risto Hilpinen. Section two contains an outline of the elements of the Hintikka-Hilpinen (hereafter H-H) system,[1] including an illustrative account of their theorem for inductive generalizations and a comparison of some of their results with Carnap's. In section three their acceptance rule for generalizations is considered. It is shown that it has a peculiar way of satisfying a consistency requirement and that it violates Salmon's criterion of linguistic variance. Finally, four objections to the H-H rule for singular hypotheses are presented.

2. ELEMENTS OF THE HINTIKKA-HILPINEN SYSTEM

H-H begin with a first-order language L_k which has k primitive monadic predicates P_i (i = 1, 2, ..., k) and N individual constants a_i(i = 1, 2, ..., N) denoting all the individuals in a given universe of discourse U. With k primitive predicates one may define $K = 2^k$ complex predicates which H-H call *"attributive constituents,* or briefly *Ct-predicates."*[2] Ct-predicates are conjunctions in which every primi-

---

[1] As developed in Jaakko Hintikka, "Induction by enumeration and induction by elimination," *The Problem of Inductive Logic*, ed. I. Lakatos (Amsterdam: North Holland Pub. Co., 1968), pp. 191–216; "Towards a theory of inductive generalization," *Proceedings of the 1964 Congress for Logic, Methodology and Philosophy of Science*, ed. Y. Bar-Hillel (Amsterdam: North-Holland Pub. Co., 1965), pp. 274–288; "A two-dimensional continuum of inductive methods," *Aspects of Inductive Logic*, ed. J. Hintikka and P. Suppes (Amsterdam: North-Holland Pub. Co., 1965), pp. 113–132. Aspects of the system are also developed by Risto Hilpinen, "Rules of acceptance and inductive logic," *Acta Philosophica Fennica*, XXII (1968), pp. 1–134.

tive predicate or its negation occurs. Using open sentences for illustrative purposes (following H-H), the jth Ct-predicate would have the form

(1) $\qquad Ct_j(x) = (\pm)P_1(x) \cdot (\pm)P_2(x) \cdot \ldots \cdot (\pm)P_k(x)$

For example, if L_k contains two primitive predicates $R(x)$ denoting 'x is a raven' and $B(x)$ denoting 'x is black' then we have the following $2^k = 4$ Ct-predicates.

$$\begin{aligned} Ct_1(x) &\equiv R(x) \cdot B(x) \\ Ct_2(x) &\equiv R(x) \cdot -B(x) \\ Ct_3(x) &\equiv -R(x) \cdot B(x) \\ Ct_4(x) &\equiv -R(x) \cdot -B(x) \end{aligned}$$

Constituents are conjunctions indicating for every Ct-predicate whether or not it is instantiated in U. They have the form
(2) $\quad (\pm)(Ex)Ct_1(x) \cdot (\pm)(Ex)Ct_2(x) \cdot \ldots \cdot (\pm)(Ex)Ct_K(x)$.
A given language L_k admits of 2^K constituents. Hence, our illustrative language L_2 has 16 constituents.

Ct-predicates   *Constituents*

	1	2	3	4	5	6	7	8	9	10	11	12	13	14	15	16
$Ct_1(x)$	0	0	0	0	1	1	1	0	0	0	1	0	1	1	1	1
$Ct_2(x)$	0	0	0	1	0	1	0	0	1	1	0	1	0	1	1	1
$Ct_3(x)$	0	0	1	0	0	0	0	1	1	0	1	1	1	0	1	1
$Ct_4(x)$	0	1	0	0	0	0	1	1	0	1	0	1	1	1	0	1
	C_0	C_1				C_2						C_3				C_4

Here the first constituent 1 is that in which no Ct-predicate is instantiated, 2–5 are those in which exactly one Ct-predicate is instantiated, and so on.

Since all general sentences of L_k are equivalent to certain disjunctions of constituents called the "distributive normal forms" of those sentences, the prior probabilities of such sentences could easily be calculated given the probabilities of each constituent. Before the latter can be assigned, values of the parameters λ and α must be selected. H-H let $\lambda(w) = w$ where w is the number of instantiated Ct-predicates conjoined in a given constituent.[3] Hence, $0 \leq w \leq K$ and for L_2

---

[2] Hilpinen, *op. cit.*, p. 51.
[3] *Ibid.*, p. 59.

there is one constituent (C_0) in which no Ct-predicate is instantiated, one (C_4) in which every Ct-predicate is instantiated, four in which exactly one is (C_1) or is not (C_3) instantiated and six in which exactly two are instantiated (C_2). With regard to α, H-H make no specific assignment. They merely note that if $\alpha = \infty$ their system yields the same values as c^* for generalizations in an infinite universe, namely, 0 and if $\alpha = 0$ the values yielded are excessively optimistic.[4] The formula used to obtain the prior probability values for the various C_w is[5]

$$(3) \quad P(C_w) = \frac{(\alpha + w - 1)!}{(w - 1)!} \bigg/ \sum_{i=0}^{K} \binom{K}{i} \frac{(\alpha + i - 1)!}{(i - 1)!}$$

The following Table 1 of probability values will provide a good picture of the effect of α on the prior probabilities of constituents.

α	C_0	C_1	C_2	C_3	C_4
0	.0619	.0619	.0619	.0619	.0619
1	.0303	.0303	.0606	.0909	.1209
2	.0088	.0186	.0531	.1062	.1858
3	.0038	.0113	.0453	.1132	.2264
4	.0019	.0077	.0387	.1161	.2708

Table I

For each row, if every $P(C_w)$ is multiplied by the number of constituents in which exactly w Ct-predicates are instantiated and these products are added, the result is approximately 1. Notice that when $\alpha = 0$, $P(C_w)$ is the same for every C_w, namely, $\frac{1}{2}^K$. Notice also that as α increases, the prior probability of C_K increases in proportion to the values of all other constituents. When $\alpha \to \infty$, $P(C_K)$ approaches 1 and the values of all other constituents approach 0.

An even better picture of the effect of α on one's probability assignments may be obtained by comparing posterior values yielded by the following derivation[6] from Bayes's formula, (3) and a likelihood function.

$$(4) \quad P(C_w, e) = 1 \bigg/ \sum_{i=0}^{K-c} \binom{K-c}{i} \frac{(\alpha + c + i - 1)!(n + w - 1)!}{(\alpha + w - 1)!(n + c + i - 1)!}$$

---
[4] Ibid., pp. 54–57.
[5] Ibid., p. 59.
[6] Ibid., p. 60.

For L_2, if e is the evidence that $n \geq 3$ different individuals have been observed "completely" and have been found to be of $c = 3$ kinds characterized by Ct_1, Ct_3 and Ct_4 (i.e., only nonblack ravens have not been observed), then the probability of the generalization 'all ravens are black' on e is just the probability of the following constituent C_c on e

C_c: $(Ex)Ct_1(x) \cdot (Ex)Ct_3(x) \cdot (Ex)Ct_4(x) \cdot (x)[Ct_1(x) v Ct_3(x) v Ct_4(x)]$.

Some representative values of $P(C_c, e)$ are given on Table 2 for $3 \leq n \leq 8$ and $0 \leq \alpha \leq 8$.

n	$c_i^*, c_{\varphi i}^*$	$\alpha=0$	$\alpha=1$	$\alpha=2$	$\alpha=4$	$\alpha=6$	$\alpha=8$
3	.571	.667	.600	.545	.462	.400	.353
4	.634	.700	.636	.583	.500	.438	.389
5	.667	.727	.667	.615	.533	.471	.421
6	.700	.750	.692	.643	.563	.500	.450
7	.727	.769	.714	.667	.588	.526	.476
8	.750	.786	.733	.688	.611	.550	.500

Table 2

Although Carnap's c* yields a value of 0 for C_c given any evidence, the instance and qualified instance confirmation of C_c on e does yield values greater than 0 and some of these have been included in Table 2 for comparison. It must not be forgotten, of course, that c_i^* and $c_{\varphi i}^*$ values are *not* the probability values of generalizations. They are merely the best substitutes available with c*. Insofar as it is legitimate to compare c_i^* and $c_{\varphi i}^*$ values with those obtained from (4), $\alpha = 1$ yields values nearest to those of c_i^* and $c_{\varphi i}^*$.

Notice that as α increases, $P(C_c, e)$ decreases for a fixed e. Hence, α may be regarded as a rough index of caution. Notice also that when $n = \alpha$, $P(C_c, e) = .500$. That is, when the number of observations compatible with our hypothesis 'all ravens are black' equals our rough index of caution, that hypothesis becomes for us an even bet. Generally, "when $n = \alpha$, all the constituents compatible with e have an equal probability $\frac{1}{2}^{K-c}$."[7] The question of evidence incompatible with a given hypothesis is not considered explicitly by H-H, but there is evidently no room for such evidence in (4). One could say that a single disconfirming instance is all that is required logically

[7] *Ibid.*

to falsify our generalization, but I suspect that a certain number of favorable instances following the single unfavorable one should be able to "restore" the probability of our hypothesis to its value prior to the unfavorable observation.

3. THE ACCEPTANCE RULE AG_n

Probabilistic acceptance rules recommend the acceptance (as true or at least as worthy of belief) of a hypothesis if and only if its probability is greater than some specified value, say $1 - \varepsilon$ where ε is some very small number like .01 or .001. H-H show that it is possible to modify such rules in order to avoid the lottery paradox.[8] The trick is to make so many observations that if the probability of a constituent is greater than $1 - \varepsilon$ after that much evidence is accumulated, *only that* constituent logically can be accepted on that evidence, i.e., roughly speaking, the trick is to unite the tactics of induction by enumeration and induction by elimination. The required number of favorable observations is n_0, the largest integer n for which

(5) $$\varepsilon' \leq \sum_{i=1}^{K-c} \binom{K-c}{i} \frac{(n+c-1)!(\alpha+c+i-1)!}{(n+c+i-1)!(\alpha+c-1)!}$$

where ε' is defined as $(1/1 - \varepsilon) - 1$.[9] Briefly, H-H recommend

"(AG_n) Accept a general hypothesis g on the basis e if and only if (i) $P(g, e) > 1 - \epsilon$, and (ii) $n > n_0$."[10]

Table 3 will give you some idea of what AG_n requires. For L_2 and our generalization that 'all ravens are black,' if $\varepsilon = .0100$ or $.0010$ then $\varepsilon' = .0101$ or $.001001$, and the largest integer n for which the right side of (5) is greater than or equal to ε' is as follows for $0 \leq \alpha \leq 5$.

---

[8] If one assumes adequacy conditions of "closure" (i.e., if h_1 and h_2 are each acceptable hypotheses then the conjunction $h_1 . h_2$ is also acceptable) and "consistency" (i.e., the set of acceptable hypotheses is logically consistent) for systems of knowledge or systems of sentences accepted as true, and if one further assumes that a high probability is sufficient for acceptability in some sense; then it is possible to have each of a set of hypotheses acceptable as well as their conjunction (by the closure requirement) *and* its negation (by the Special Addition Principle of the probability calculus). This is the lottery paradox.
[9] Hilpinen, *loc. cit.*, p. 62.
[10] *Ibid.*, p. 64.

	$\varepsilon' = .0101$	$\varepsilon' = .001001$
α	n	n
0	294	2994
1	393	3993
2	492	4992
3	591	5991
4	690	6990
5	789	7989

Table 3

Thus, 'all ravens are black' is acceptable in L_2 if and only if its probability is, say, greater than .99, 789 favorable and no unfavorable observations have been made, and $\alpha = 5$. If $\alpha = 4$ then only 690 favorable observations are required. Confronted by such mechanically produced numbers as 789 and 690, I must admit that it is extremely difficult to "warm up to the idea" that these numbers and only these suffice in the given circumstances. One has the suspicion that there are some hypotheses (or hypotheses with certain predicates, if you like) such that the idea of trotting out, say, 789 compatible instances in order to guarantee their acceptability is somehow beside the point. As Hume and others have observed, some hypotheses seem to be warranted after a single favorable observation and others seem to be unwarranted given any number of favorable observations. It may be the case that *no* explication of induction by enumeration can accommodate (i.e., justify) such inferences, but if so, this must be regarded as a good reason for turning one's attention to other procedures.

Again, it must be emphasized that it is logically or mathematically impossible for any constituent but C_c or a generalization in which C_c occurs in its distributive normal form to be acceptable on e once both conditions of AG_n have been met.[11] This of course implies that any set of acceptable hypotheses will satisfy the adequacy requirement of consistency, but it seems to be a peculiar way to satisfy that requirement. After all, what H-H are telling us is that two generalizations g_1 and g_2 may be accepted if and only if there is a single acceptable constituent C_c occurring in each generalization *in virtue of which* g_1 and g_2 are acceptable. Clearly then, if the content of C_c exhausts g_1 and g_2, although *prima facie* we have two hypotheses, we really have only one. Hence, it is misleading to describe g_1 and g_2 as *two*

[11] *Ibid.*, p. 63.

acceptable and mutually consistent hypotheses. On the other hand, if the content of C_c does not exhaust g_1 and g_2 then some part of the latter claims must be unwarranted, since strictly speaking only C_c is warranted. More precisely, g_1 and g_2 will be disjunctions of constituents all of which are unwarranted but C_c. Unwarranted constituents are such that one should either suspend judgment about them or reject them. If the unwarranted disjuncts of g_1 and g_2 are rejected, then we are again left with a single acceptable hypothesis, C_c. If we merely suspend judgment on these constituents, then they are still not accepted. Hence, C_c is still our sole accepted hypothesis. Thus, it is again misleading to describe g_1 and g_2 as *two* acceptable and mutually consistent hypotheses. In fact, our entire corpus of acceptable generalizations always consists of a single acceptable constituent, if it consists of anything at all. Therefore, whether the content of an acceptable constituent exhausts that of all the hypotheses accepted on a given body of evidence or not, it does not seem to me that H-H's procedure satisfactorily fulfills the requirement of consistency for the set of acceptable generalizations.

Supposing that one is prepared to live with the peculiar satisfaction of the consistency requirement we have just described, one may still balk at the prospect of adopting a system of inductive logic that violates Salmon's criterion of linguistic invariance (CLI).[12]

The following example shows that the H-H system does violate CLI. Reinterpret the predicates of L_2 such that $R(x)$ designates 'x is a Roman' and $B(x)$ designates 'x is a bachelor.' Then, according to Table 2, if $\alpha = 0$ and $n = 5$ favorable instances have been observed, the probability of the generalization

g_1: All Romans are bachelors.

is $P(g_1, e) = .727$. Now, consider a different language L_3 which contains the predicates $R(x)$, $U(x)$ for 'x is unmarried' and $M(x)$ for 'x is male.' Hence, there are $K = 2^k = 8$ Ct-predicates in L_3, namely,

$Ct_1(x) \equiv R(x) \cdot U(x) \cdot M(x)$ $\qquad Ct_5(x) \equiv -R(x) \cdot U(x) \cdot M(x)$
$Ct_2(x) \equiv R(x) \cdot U(x) \cdot -M(x)$ $\qquad Ct_6(x) \equiv -R(x) \cdot U(x) \cdot -M(x)$
$Ct_3(x) \equiv R(x) \cdot -U(x) \cdot M(x)$ $\qquad Ct_7(x) \equiv -R(x) \cdot -U(x) \cdot M(x)$
$Ct_4(x) \equiv R(x) \cdot -U(x) \cdot -M(x)$ $\qquad Ct_8(x) \equiv -R(x) \cdot -U(x) \cdot -M(x)$

The generalization whose probability we wish to determine is

g_2: All Romans are unmarried males.

[12] See Chapter VI, Section 5.

which is obviously equivalent to g_1. Our evidence e is that $n = 5$ favorable instances have been observed, in particular, individuals with $Ct_1(x)$ and $Ct_5(x) - Ct_8(x)$. So, only the three possible unfavorable cases have not been observed. If $\alpha = 0$ then $P(g_2, e) = .287$, which is considerably less than the .727 obtained for the equivalent generalization on the same evidence. Therefore, the H-H system clearly violates CLI. Furthermore, this violation has occurred when the "weight" of the logical factor on our probability values is at its minimum, i.e., $\alpha = 0$! Of course, if one could carry through the program suggested by Lehrer, violations of CLI might be regarded as innocuous.[13] But I do not know what H-H's view about this program is, or what else they may want to do about such problems. One thing is certain, however, and that is that one of the objections Hilpinen raises against Isaac Levi's acceptance rule[14] is that it violates an alleged adequacy condition which has virtually the same significance as CLI.

4. THE ACCEPTANCE RULE AS

It is a consequence of H-H's theory and apparently acceptable to them that

"(AS) A singular hypothesis '$A(a_i)$' is acceptable on the basis of e if and only if '$(x)A(x)$' is (on the basis of e) an acceptable generalization."[15]

We have already seen that c_i^* and c_{qi}^* violates AS, but one can hardly use inherently unacceptable formulas to criticize an adequacy condition. I believe, however, that there are at least four good reasons for rejecting AS and any theory (such as H-H's) which entails it. These are illustrated in the following *kinds* of counter-instances to AS.

Case 1. Suppose we have a device, whether manmade or natural, that has regularly turned out two kinds of events for the past n trials. Letting K_1 and K_2 represent these kinds, we have observed e, a sequence of occurrenses representable by

$$K_1 K_2 K_1 K_2 \ldots K_n.$$

Presumably the larger n becomes, the greater one's confidence in the

---

[13] *Ibid.*, Sections 6 and 7.
[14] Hilpinen, *loc. cit.*, pp. 100–101.
[15] *Ibid.*, p. 74. It is also worthwhile to compare L. J. Cohen's analogue of this rule in Alex C. Michalos, "An alleged condition of evidential support," *Mind*, 78, (1969), pp. 440–441.

next occurrence ought to become. If the last event was K_1, the next event will be K_2 and vice versa with a probability approaching 1 as n approaches infinity. Nevertheless, the probability that all events will be K_1 or that they will all be K_2 is apparently 0.

Case 2. Suppose e is the information that Smith has been holding his breath for 80 seconds. Then it would seem highly probable that he can hold his breath for the 81st second, but impossible that he can hold his breath forever, i.e., for all seconds beyond the first 80.

Case 3. Suppose e is the information that Smith has resolved to have exactly three drinks whenever he attends a party, that he has always fulfilled his resolution, and that he has already had two drinks at this party. Then the probability that he will take the next drink offered to him might well be very high although the probability that he will take every drink offered to him might be very low.

Case 4. Suppose e is the information that Smith has been away from Jones for several years and that they are old and dear friends. Then the probability that Smith will do whatever Jones feels like doing the next time they meet might be very high, while the probability that Smith will do whatever Jones feels like doing every time they meet in the future might well be low.

Insofar as AS prohibits the kinds of probability appraisals made in these four cases, it seems to be an unacceptable adequacy condition for a system of inductive logic. Insofar as H-H's system entails AS, it too must be regarded as unacceptable. Granted that it is still in an embryonic form, it does not seem to be headed in the right direction if it has the unsatisfactory consequences cited here.

CHAPTER IX

COST-BENEFIT VERSUS EXPECTED UTILITY ACCEPTANCE RULES

1. INTRODUCTION

In Chapter II it was noted that Carnap's Rule of Maximizing Estimated Utility may be used as a measure of acceptability in virtually *any* sense of this difficult term. By including different kinds of considerations with different amounts of importance in the determination of utility values, the scope of application of the rule tends to be exhaustive. As a matter of fact, which Carnap has himself emphasized, the rule he recommends as a normative principle is very old. It was invented by Daniel Bernoulli and modified by Thomas Bayes. Hence, it is sometimes referred to as Bernoulli's rule and sometimes as Bayes's rule, and, in the present case, as the Bernoulli-Bayes rule.

A number of influential philosophers besides Carnap have recommended the Bernoulli-Bayes rule or some variation of it as a first approximation or step in the right direction toward a solution of the problem of providing a criterion, principle or rule for determining the acceptability of scientific hypotheses. But, so far as I know, no one has suggested that some sort of benefits-less-costs rule might be more advantageous, and it is roughly this idea that I wish to explore and ultimately vindicate. More precisely, I shall attempt to prove the *normative* claim that a cost and benefit dominance principle of acceptance ought to be preferred to any sort of Bernoulli-Bayes principle because right now and for the foreseeable future the former performs better and cannot perform worse than the latter (in a sense of 'perform' that will be elucidated below).[1]

---

[1] Less thorough comparisons of these and similar rules with respect to different applications may be found in K. R. MacCrimmon, *Decisionmaking Among Multiple-Attribute Alternatives: A Survey and Consolidated Approach*, Memorandum RM-4823-ARPA, The RAND Corporation, Santa Monica, 1968 and P. M. Pruzan, "Is cost-benefit analysis consistent with the maximization of expected utility?" *Operational Research and the Social*

Although most of the chapter consists of a detailed analysis and comparison of the two relevant principles, their requirements and applications, I shall begin with a brief outline of the basic elements of each in sections two and three in order to provide a general orientation and more or less common background for our discussion. In section four relevant senses of the terms 'preferable,' 'performs better,' 'effective,' 'efficiency,' and 'degree of sophistication of information' are explained. These are necessary for the comparison of the Bernouilli-Bayes rule with that of the variant of a benefits-less-costs rule in sections five through seven. Section six contains a detailed examination of William Harvey's implicit use of his own variant of a benefits-less-costs rule. In the eighth section strategies for increasing the effectiveness of the rule recommended here are presented and the final brief section suggests areas for future research.

2. MAXIMIZATION OF EXPECTED UTILITY

Proponents of the rule enjoining the maximization of expected utility, which we shall hereafter abbreviate as MEU, imagine a decision-maker confronted with a set of (practically speaking) mutually exclusive and exhaustive possible courses of action from which one that is optimal must be adopted. The decision-maker knows that the payoff or *utility* (in some sense of this word which will be explained later) that he obtains from his choice will be partially determined by events which are (practically speaking) mutually exclusive, exhaustive and beyond his control. If he has objective probability values (i.e., relative frequencies, propensities, physical range measures, etc.) for the occurrence of these events then he is operating under conditions of risk. If he does not have such values then he is in a situation of uncertainty, but he will transform it into a situation of risk by determining appropriate subjective probability values (i.e., betting quotients, degrees of belief, etc.) for the events. Given all this data he is ready to use MEU, which, as a normative principle, prescribes the acceptance of that course of action whose sum of probability-weighted utilities is larger than that of any of its alternatives. If 'U_i' and 'p_i'

Sciences, ed. J. R. Lawrence (London: Tavistock Pub., 1966), pp. 319–336. A defense of the Bernoulli-Bayes rule for "multi-attribute" problems may be found in H. Raiffa, *Preferences for Multi-Attributed Alternatives*, Memorandum RM-5868-DOT/RC, The RAND Corporation, Santa Monica, 1969. An excellent survey of recent work on cost-benefit analysis may be found in A. R. Prest and R. Turvey, "Cost-benefit analysis: a survey," *The Economic Journal*, 75 (1965), pp. 683–735.

represent the utility and probability values, respectively, of the ith of n payoffs obtainable by adopting some hypothesis, then

(1) $$\Sigma\, p_i U_i\ (i = 1, 2, \ldots, n)$$
or
(2) $$p_1 U_1 + p_2 U_2 + \ldots + p_n U_n$$

represents the expected utility of accepting that hypothesis. Clearly, when there is no risk involved then formulae (1) and (2) shrink to the single utility value whose procurement is certain (has a probability value of unity) given the acceptance of that hypothesis. Following MEU then, one would simply accept that hypothesis whose expected-utility (or utility in the limiting case) was greater than that of its alternatives. If two or more hypotheses have the same expected-utility and it is higher than those of the alternatives, then each of the former should be regarded as equally acceptable.

For our purposes we may think of the decision-maker described above as a practicing theoretical or applied scientist, and we may assume that his possible courses of action are the adoption of certain hypotheses as bases for further action. Although one *can*, as Isaac Levi has shown us,[2] imagine a person accepting a hypothesis not as a basis for action but as somehow suitable for admission into his total corpus of knowledge, our concern here is with the provision of hypotheses more or less directly related to action.[3] If we regard a hypothesis as acceptable then, at the very least, it merits the investment of further resources such as research facilities and activity, time, money, energy, and so on. Note, however, that our concern with the acceptability of hypotheses as bases for action does not alter any of the formal or logical aspects of the problem. Only the content of the utility function or the arguments to be included in that function would be different for Levi's decision-makers and mine.

3. COST-BENEFIT DOMINANCE

Our exposition of a variation of a benefits-less-costs rule, which we will call 'the principle of cost-benefit dominance' and abbreviate CBD, may begin with a decision-maker in roughly the same situation we described for MEU. He is confronted with a similar set of (practi-

---

[2] I. Levi, "On the seriousness of mistakes," *Philosophy of Science*, 29 (1962), pp. 47–65.
[3] Compare R. M. Chisholm, "Lewis' ethics of belief," *The Philosophy of C. I. Lewis*, ed. P. A. Schilpp, LaSalle, Illinois: Open Court Press, (1968), pp. 223–242.

cally speaking) mutually exclusive and exhaustive hypotheses and events. He may be able to assign some sort of probability value to the occurrence of each event and he may not. In any case, such values are not required. Similarly, utility values are not required. Instead of assigning a utility value to the payoffs he will receive as a result of this or that combination of accepted hypothesis and turn of events, he merely determines the "raw forms" of the benefits and costs attached to each combination. For example, instead of noting that a hypothesis adequately accounts for a certain phenomenon provided that certain events take place rather than some others, coheres with a well-established theory in another domain given those events and has a Reichenbachian weight of .7, *and* that all of this gives it a utility value of .8, he merely lists its benefits in their "raw form" (i.e,. it coheres provided that such and such is the case, etc.). Similarly, he lists its costs and the benefits and costs for its alternatives given the various contingencies. It must be assumed, of course, that he is able to weakly order the "raw form" data *within* each attribute and the corresponding preferences. For example, he must be able to determine whether two hypotheses are equally explanatory or one explains more than the other; he must be able to rank order any three distinct levels of explanatory power transitively; and he must recognize, that his preferences ought to be perfectly positively correlated with the "raw form" data, e.g., he ought to prefer a hypothesis that explains more phenomena to one that explains less (if all other things are equal). It does not have to be assumed that he is able to weakly order the "raw form" data across attributes, e.g., he never has to be able to rank order different levels of explanatory power, coherence, testability, etc. on a single scale of some sort. In other words, he is obliged to make intra-attribute comparisons but not inter-attribute comparisons.

The net result of this analysis is a battery of matrices (i.e., a different matrix for each attribute) which constitute a comparative "profile" of each hypothesis with respect to its alternatives. Schematically, the matrices in such a battery for the attributes of, say, simplicity, explanatory power, precision, coherence, testability, etc. would each look like this.

Possible Events (States of Nature)

$E_1 \quad E_2, \ldots, E_n$

$H_1 \; p_1 B_1^1 \; p_2 B_2^1, \ldots, p_n B_n^1$

Hypotheses
$$H_2 \quad p_1B_1^2 \quad p_2B_2^2, \ldots, p_nB_n^2$$
$$\vdots \quad \vdots \quad \vdots \quad \vdots$$
$$H_n \quad p_1B_1^n \quad p_2B_2^n, \ldots, p_nB_n^n$$

The entry for row H_1 column E_2, for example, would tell us that if the hypothesis represented by 'H_1' is accepted and the event(s) represented by 'E_2' occur, then we will obtain a benefit represented by 'B_2^1' with a probability represented by 'p_2.' Since 'p_2' and 'B_2^1' need not represent numerical values of any sort, the juxtaposition of these two signs must not be taken to mean multiplication (as in formulae (1) and (2)). The superscript on 'B_2^1' indicates the hypothesis 'H_1' and the subscript indicates the event(s) 'E_2.' The 'B' is short for 'benefits.' They might be a high degree of explanatory power, simplicity, coherence with other theories, precision, etc. While various philosophers and scientists[4] besides Popper and Carnap have made recommendations as to which attributes ought to be included in an optimal set, all that is assumed here is that such a set would contain, say, more than a couple and less than a couple dozen members, none of which would have to be entirely independent (in any sense) of the others. In matrices for such costs as required set-up time, computational effort, special facilities, technical assistance, money, operationalization, etc. the 'B' would be replaced by a 'C' for 'costs.'[5] As with formulae (1) and (2), when there is no risk involved then such matrices shrink to a single item, namely, a column indicating the various benefits (costs) that will be obtained (borne) with certainty given the acceptance of a particular hypothesis.

If for every possible contingency and for every attribute, the benefits and costs of accepting one hypothesis are preferable (i.e., ought to be preferred) to those of accepting another then the former *strongly dominates* the latter. If for some contingency and for some attribute, the benefits and costs of accepting one hypothesis are preferable to those of another *and* for all of the remaining contingencies and

---

[4] See, for example, G. Buchdahl, *Metaphysics and the Philosophy of Science*, (Oxford: Basil Blackwell, 1969); M. Bunge, "The weight of simplicity in the construction and assaying of scientific theories," *Philosophy of Science*, 28 (1961), pp. 120–149; I. J. Good, "Corroboration, explanation, evolving probability, simplicity and a sharpened razor," *B.J.P.S.*, 19 (1968), pp. 123–143; L. Laudan, "Theories of scientific method from Plato to Mach," *History of Science*, 6 (1968), pp. 1–63 and H. Margenau, *The Nature of Physical Reality*, (New York: McGraw-Hill, 1950).

[5] A fairly thorough analysis of decision-making costs may be found in Alex C. Michalos, "The costs of decision-making," *Public Choice*, 9 (1970).

attributes the benefits and costs of accepting the latter are not preferable to those of the former (i.e., some are exactly alike and others are less preferable), then the former dominates the latter.[6] Hence, if one hypothesis strongly dominates another then the former also dominates the latter, but the converse is not true. According to CBD then, the hypothesis that ought to be accepted is the one which dominates all of its alternatives. If two or more hypotheses have the same benefits and costs but dominate all others then they should be regarded as equally acceptable. In general, depending on the particular benefits and costs involved, research in a given problem area should continue until some hypothesis emerges as dominant overall of its alternatives.

4. PREFERABILITY AND SUPERIOR PERFORMANCE

I am assuming that one principle of acceptance is *preferable* to another provided that the former may perform better and cannot perform worse than the latter. Moreover, one principle *performs better* than another if and only if it is more effective and more efficient than the other. An acceptance principle is *effective* exactly insofar as it is possible in every sense of this term to isolate, identify or select acceptable hypotheses by applying it. If it were impossible in any sense to apply a principle then it could not be applied and, consequently, could not identify anything. So it would be completely ineffective. The *efficiency* of an acceptance principle may be defined and measured by the ratio of its effectiveness to the number of assumptions and the degree of sophistication of the kinds of information required for its application.[7] The *degree of sophistication* of a particular piece of information may be determined by the sorts of scales (nominal, ordinal, interval or ratio) or concepts (qualitative, comparative or quantitative) required to accurately express that information, e.g., 'this is hot' may be regarded as a less sophisticated piece of information than 'this is hotter than that'

---

[6] These two definitions, of course, are merely special applications of the famous Pareto Principle that has been used widely by economists since Vilfredo Pareto's *Cours d'Economie politique* (1877), e.g., in K. J. Arrow, *Social Choice and Individual Values*, (New York: John Wiley and Sons, 1951); J. C. Harsanyi, "Cardinal welfare, individualistic ethics and interpersonal comparisons of utility," *Journal of Political Economy*, 63 (1955), pp. 309–321; C. Hildreth, "Alternative conditions for social orderings," *Econometrica*, 21 (1953), pp. 81–91 and J. Rothenberg, *The Measurement of Social Welfare*, (Englewood Cliffs, New Jersey: Prentice-Hall, Inc., 1961).

[7] "Effectiveness" and "efficiency" are analyzed in greater detail in Alex C. Michalos,, "Efficiency and morality," a paper read at the Annual Meeting of the Western Division of the American Philosophical Association, May 1970.

which is less sophisticated than 'this has a temperature of 90°F.' Hence, if it could be shown that CBD and MEU are equally effective or that CBD is more effective than MEU *but* that CBD requires fewer assumptions and/or less sophisticated information than MEU, then the superior efficiency of CBD would be established. That, of course, would establish its superiority of performance and, therefore, its preferability over MEU, which is my central thesis. In the next two sections I shall attempt to establish this thesis roughly as follows. In section five it will be shown that MEU requires information of a more sophisticated sort and, consequently, more assumptions than CBD. Some of this information is not now nor will it be in the foreseeable future available. It follows then, that MEU is now and will be for some time to come *completely ineffective*. In section six I review Harvey's defense of his theory of the motion of the heart and blood against that of Galen as a paradigm case study of an undoubtedly successful defense of a scientific hypothesis *and* of an implicit application of CBD. Any adequate acceptance rule would have to disclose the superiority of Harvey's view over Galen's and, as a matter of fact, it is fairly apparent that something like CBD was behind Harvey's presentation of the evidence for his view. By means of this historical example then, the *effectiveness* of CBD is established. Thus, because CBD is more effective and must have a higher efficiency ratio than MEU, the former performs better and cannot perform worse than MEU for the present and foreseeable future. Hence, the superiority and preferability of CBD over MEU is established.

5. MEU VERSUS CBD: COMPARISON OF REQUIREMENTS

The requirements and general prospects of our two principles may be thoroughly compared in five respects.

5.1 In the first place it is apparent that MEU does but CBD does not require numerical probability values. With MEU it is not enough to be able to determine that a certain event and concomitant attribute value is probable, very probable, improbable, more probable than some other, as probable as some other, etc. Such frequently useful qualitative and comparative probabilistic judgments are worthless for MEU, because the latter can only "process" quantitative judgments; e.g., judgments of the form 'The degree of probability of obtaining a value of x for attribute A in the event that E is r.' Thus, one who uses MEU must assume that all of the notoriously difficult philosophic problems

COST-BENEFIT VERSUS UTILITY ACCEPTANCE RULES 99

involved in obtaining initial numerical probability values (for the particular procedures employed) have been satisfactorily solved.[8] Moreover, granted that a given position is philosophically unobjectionable in itself, it must still be assumed that every contingency that might be relevant to the acceptance of any hypothesis can be meaningfully assigned a numerical value by that procedure, i.e., that in one way or another it is always meaningful to transform conditions of uncertainty into conditions of risk. As Ellsberg has shown,[9] however, proponents of minimax, maximin and maximax decision rules cannot and would not accept this assumption. On the other hand, both the assumption and its denial are irrelevant to CBD, for the latter does not require probability values of any sort but it can always use them when they are available.

5.2 Just as CBD can but MEU cannot get along without numerical probability values, the former can and the latter cannot proceed without numerical utility values. There seem to be two *prima facie* possible ways to produce these values, the first of which will be shown to be abortive and the second of which is largely wishful thinking. The former will be considered in this subsection and the latter in the next.

To begin, it should be noted that the numerical utility values required are cardinal and not merely ordinal. The former are necessary because they are the only ones that can be meaningfully added or multiplied, and both of these operations must be performed on the utility values used by MEU. So, some means of obtaining at least an interval scale of numerical utility values must be found.[10]

One way to tackle the problem of securing an interval scale of utility values is to try to generate them from a simple rank ordering of attribute values. The most popular and manageable means to this end is the standard-gamble technique of von Neumann and Morgen-

---

[8] See, for example, Alex C. Michalos, *Principles of Logic*, (Englewood Cliffs: Prentice-Hall, Inc., 1969), Chapter Five; S. F. Barker, *Induction and Hypothesis*, (Ithaca, New York: Cornell University Press, 1957), Chapters Three and Four; and W. C. Salmon, *The Foundations of Scientific Inference*, (Pittsburgh: University of Pittsburgh Press, 1966), Chapters Four and Five.

[9] D. Ellsberg, *Risk, Ambiguity and the Savage Axioms*, P-2173, The RAND Corporation, Santa Monica, 1961. See also A. W. Burks, "The pragmatic-Humean theory of probability and Lewis' theory," *The Philosophy of C. I. Lewis*, ed. P. A. Schilpp, (LaSalle, Illinois: Open Court Press, 1968), pp. 415–464, and J. Milnor, "Games against nature," *Decision Processes*, ed. R. M. Thrall, C. H. Coombs and R. L. Davis, (New York: John Wiley and Sons, 1960), pp. 49–60.

[10] On the problem of scales and their transformations see R. L. Ackoff, *Scientific Method*, (New York: John Wiley and Sons, 1962), Chapter Six; and P. C. Fishburn, *Decision and Value Theory*, (New York: John Wiley and Sons, 1964), Chapter Four.

stern.[11] Unfortunately (for defenders of MEU), while this technique can (in principle though not always in fact[12]) be used to produce an interval scale of utility, the utility involved is not the sort that is of interest to philosophers of science. The latter are concerned with "epistemic utilities." In Hempel's words, "*epistemic utilities* ... represent 'gains' and 'losses' as judged by reference to the objectives of 'pure' or 'basic' scientific research; in contradistinction to ... *pragmatic utilities*, which would represent gains or losses in income, prestige, intellectual or moral satisfaction ...."[13] And Levi writes "when an investigator declares himself to be engaged in an effort to replace agnosticism by true belief ... there is no need to ascertain his 'true feelings' ...."[14]

Presumably the fundamental logical distinction between a scale of "epistemic utility" and "pragmatic (including psychological) utility" is that only the former could have some normative force for a scientist *as* a scientist.[15] Moreover, it might be thought that by making certain adjustments in the von Neumann-Morgenstern technique, a scale with normative force could be constructed. Indeed, this idea seems to be behind Levi's efforts in *Gambling With Truth*.[16] It is easy to demonstrate, however, that the von Neumann-Morgenstern technique cannot yield the required scale. According to that technique, a decision-maker begins by rank ordering attribute values. To keep things simple, suppose he has a single attribute, say, explanatory power, and can distinguish in a publically observable or intersubjectively testable fashion, three ranks, low, average and high. The latter may be represented by 'L,' 'M' and 'H', respectively. Clearly, he prefers

H to M to L.

---

[11] J. von Newmann and O. Morgenstern, *The Theory of Games and Economic Behavior*, (Princeton: Princeton University Press, 1947).

[12] Objections are raised in Alex C. Michalos, "Postulates of rational preference," *Philosophy of Science*, 34 (1967), pp. 18–22 and "Estimated utility and corroboration," *B.J.P.S.*, 16 (1966) pp. 327–331.

[13] C. G. Hempel, "Deductive-nomological versus statistical explanation," *Minnesota Studies in the Philosophy of Science*, Vol. III, ed. H. Feigl and G. Maxwell, (Minneapolis: University of Minnesota Press, 1962), p. 156.

[14] I. Levy, *Gambling With Truth*, (New York: Alfred A. Knopf, 1967), p. 76.

[15] A vast amount of literature has been produced by proponents and opponents of "pragmatic utility," and it is doubtful that I could contribute anything novel to the discussion here. Interested readers may find critiques of the concept in: Arrow, *op. cit.*, Ellsberg, *op. cit.*; MacCrimmon, *op. cit.*; Michalos, "Estimated utility and corroboration," *loc. cit.*; Michalos, "Postulates of rational preference," *loc. cit.*; Rothenberg, *op. cit.*; and A. Charnes and A. C. Stedry, "The attainment of organization goals through appropriate selection of subunit goals," *Operational Research and the Social Sciences*, ed. J. R. Lawrence, (London: Tavistock Pub., 1966), pp. 147–164.

[16] Levi, *loc. cit.*, p. 50.

He then assigns a value of 0 to the lowest rank and 1 to the highest rank. To determine the numerical value of the remaining rank, he considers various choices (gambles) that might be put to him having the form

$$M \text{ versus } pL + (1 - p)H$$

where 'pL' is short for 'obtaining a low ranking hypothesis with a probability value of p.' Clearly, if $p = 0$ then the right side of the gamble would be preferred (because the choice is then between M and H, and H is preferred to M), and if $p = 1$ then the left side is preferred (because M is preferred to L). By varying the value of p appropriately, the decision-maker can (often if not always) reach a point where both sides seem equally attractive. At that point he merely computes the numerical value of the right side and that tells him the value of the left side, i.e., the value of the remaining rank.

Now, in order to put this whole procedure to work *normatively*, one would have to have some criterion, rule or principle that prescribes the appropriate probability values. For the case before us, it might have the form: Given a gamble between 'M' and 'pL + (1 − p)H' a decision maker (with his eye on the "gains" and "losses" to "pure" science) ought to reach a state of indifference when and only when $p = r$. The only way to justify such a prescription, however, is to show that M must have a certain "epistemic utility" value, and that can only be shown *after* the interval scale of "epistemic utility" values for explanatory power has been constructed. That, of course, is much too late for anyone who thought the standard-gamble technique would be useful. It is *not* useful, because one must already possess the very information it is supposed to, but can only redundantly provide. Hence, it is impossible to use this technique to construct a scale of "epistemic utility" values which would have normative force for a scientist *as* a scientist. Therefore, the first *prima facie* possible way of producing the scale required by MEU has been shown to be abortive.

5.3 The other *prima facie* plausible way to obtain the required scale is to construct interval scales of measurement for *every* relevant attribute and then transform the attribute values into utility values. Mathematically, at least, the transformations would be relatively straightforward. As a matter of fact, both Hempel and Levi have begun their constructions of "epistemic utility" measures from approximately this point.[17] Both of them have worked primarily *from* a

[17] See Hempel, "Deductive-nomological versus statistical explanation," *loc. cit.*, p. 154 and Levi, *Gambling With Truth*, p. 71.

"content measure" which yields values unique up to a linear transformation *to* a measure of "epistemic utility" which yields similar values. While there is nothing objectionable about this procedure in itself, it must be emphasized that it is at best a first step on a very long journey. Even if we had an acceptable measure of content, we would still almost certainly need measures of simplicity, explanatory power, precision of predictions, coherence with theories in other domains, probability, etc. in order to apply MEU. Needless to say, such measures are not available now and, judging from the history and present status of the unresolved issues surrounding the measurement of probability, they will not be availalbe in the foreseeable future either. Again, however, such measures (interval scales) are not necessary for CBD, which may be applied as soon as one is able to construct a simple rank ordering of attribute values.

5.4 As has been suggested, the transformation of one interval scale into another is mathematically straightforward. If 'u' and 'a' represent the "epistemic utility" and attribute values, respectively, of accepting a certain hypothesis then

$$u = f(a)$$

where 'f' represents any transformation function up to and including the linear one

$$u = ma + n$$

where 'm' and 'n' represent constants and $m \neq 0$. The extremely difficult problems before proponents of MEU with respect to these transformations are not, therefore, mathematical or merely technological. They are plainly philosophical. In particular, they involve the appropriate selection of 'f,' or, in the linear case, the values of the constants 'm' and 'n.' In the latter case, for example, given interval scales of measurement for every relevant attribute, in order to apply MEU one must be able to systematically choose the appropriate constants and to present a plausible justification of his choices. The burden of proof then, that such and such a value for a certain attribute is worth this or that much "epistemic utility" falls squarely upon MEU's adherents, and it is a burden which can hardly fail to create interminable haggling. After all, what principles could one invoke to show that, say, a hypothesis with a degree of simplicity of .8 ought to be assigned an "epistimic utility" of .8 or .6 or anything else? What principles could be used to prove that, say, a coherence value of .8 ought to be worth more or less than a simplicity value of .8?

What could we use as a basis of comparing the significance of various attributes in order to transform their values into "epistemic utility" values? Clearly some sort of interattribute comparisons will be required to justify the transformation functions, but there is no basis for such comparisons. Of course, if we already knew the "epistemic utility" values corresponding to every attribute value for all attributes, then such comparisons would be self-evident. But that is beside the point. It is precisely those utility values that we are unable to obtain, because the required principles and bases for comparison do not exist. Indeed, even the idea that they (not to mention their justifications) might be forthcoming in the near future seems farfetched to say the least. Thus, we have another good reason for expecting perennial ineffectiveness from MEU and for turning to CBD instead.

5.5 Supposing, for the sake of argument, that all of the aforementioned problems were satisfactorily solved, proponents of MEU would still be faced with an amalgamation problem, i.e., with a problem of combining all of the individual "epistemic utility" values into a single most representative or appropriate value. Some rule of combination must be constructed and its plausibility defended, and this just creates more unnecessary work in view of the availability of CBD. Because a number of amalgamation rules have already been developed, however, this final step may be the least troublesome of all.[18] But it is still a piece of excess baggage.

To summarize the arguments in this section, I have been attempting to establish the ineffectiveness of MEU on the ground that it requires a more sophisticated sort of information than is now or will be in the foreseeable future available. To obtain this information, a number of problematic assumptions have yet to be made and substantiated. CBD, on the other hand, does not require such sophisticated information or, consequently, its attendant assumptions. Thus, a *prima facie* case for CBD over MEU has been established. But this is not enough. It is one thing to show that CBD does not have the infelicities of MEU and another to show that CBD is effective. It is to the latter task that we shall now turn. If it can be shown that CBD is at all effective now or can be expected to be effective in the foreseeable future, then both its superior effectiveness and efficiency over MEU will be established.

---

[18] See, for example, Harsanyi, *op. cit.*, Hildreth, *op. cit.*; N. Rescher, *Introduction to Value Theory*, (Englewood Cliffs: Prentice-Hall, Inc., 1969), Chapter Eight, and Rothenberg, *op. cit.*

6. HARVEY'S IMPLICIT USE OF CBD

In Chapter 14 of his classic *Anatomical Disquisition on the Motion of the Heart and Blood in Animals*[19] Harvey summarizes his position as follows.

> Since *all things*, both argument and ocular demonstration, show that the blood passes through the lungs, and heart by the force of the ventricles, and is sent for distribution to all parts of the body, where it makes its way into the veins and porosities of the flesh, and then flows by the veins from the circumference on every side to the centre, from the lesser to the greater veins, and is by them finally discharged into the vena cava and right auricle of the heart, and this in such a quantity or in such a flux and reflux thither by the arteries, hither by the veins, *as cannot possibly be supplied by the ingesta* and *is much greater than can be required for mere purposes of nutrition*; it is absolutely necessary to conclude that the blood in the animal body is impelled in a circle, and is in a state of ceaseless motion; that this is the act or function which the heart performs by means of its pulse; and that it is the sole and only end of the motion and contraction of the heart.[20] (Italics added)

This famous paragraph is instructive in a number of ways, four of of which are relevant to our discussion.

First, the fact that Harvey insists that "all things" show that his position is sound reveals an implicit acceptance of the methodological rule we have been adocating, namely, CBD. He obviously assumes that the sort of complete dominance that his hypothesis (theory) has over its rivals is a sufficient reason for accepting it. Some of the details of that dominance will be described below.

Second, the long conjunction from "show" to the italicized phrases outlines the circular path taken by the blood through a body. This is illustrated in Figure 1 beside its most widely held alternative which was developed by Galen.[21] We shall have more to say about these diagrams below, but for now it is enough to notice that beginning with the vena cava, the arrows indicating the direction of flow in Harvey's

---

[19] W. Harvey, *An Anatomical Disquisition on the Motion of the Heart and Blood in Animals* Willis's translation revised and edited by A. Bowie, London, 1889 and reprinted in *Classics of Medicine and Surgery*, ed. C. M. B. Camac, (New York: Dover Pub., 1959), pp. 33–111.

[20] *Ibid.*, p. 93.

[21] The path outlined here for Galen's view has been put together from excerpts from Galen's *On the Functions of Parts of the Human Body* in A. C. Crombie, *Medieval and Early Modern Science. Vol. I*, (Garden City, New York: Doubleday and Co., 1959), pp. 162–167; D. Fleming, "Galen on the motions of the blood in the heart and lungs," *Isis*, 46(1955), pp. 14–21; M. Graubard, *Circulation and Respiration*, (New York: Harcourt, Brace and World, Inc., 1964), pp. 41–57; C. Singer, *A Short History of Anatomy and Physiology from the Greeks to Harvey*, (New York: Dover Pub., 1957), pp. 58–62, and from the remarks of the historians themselves

COST-BENEFIT VERSUS UTILITY ACCEPTANCE RULES 105

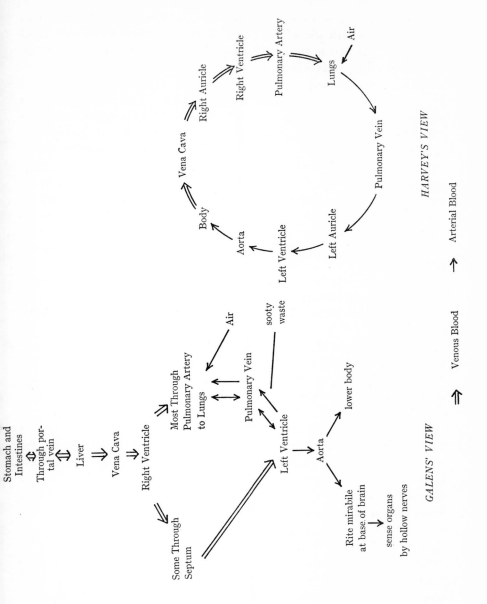

view form a circle while those in Galen's view proceed along two straight paths with some "back-up." This illustrates the fundamental discrepancy between the view that Harvey is advocating and the alternative that he is rejecting.

Third, the italicized phrases refer to Galen's theory that the blood is produced from ingesta and dispersed outward continuously from the liver to the rest of the body for nutrition. Although Cesalpino[22] had somewhat vaguely and inconsistently advocated the theory that there was a daily ebbing and flowing (like the tides) between the heart and the veins and arteries, and although there was some two-way movement in Galen's view certainly in the pulmonary and portal veins and possibly throughout the system as a result of his theory of "attractive" and "expulsive faculties" (explained below), Harvey's primary target is the idea that the blood flows more or less continuously from the liver. It is this hypothesis especially that his continuous circulation theory is to replace.

Fourth, the remarks following the italicized phrases emphasize the main aspects of his theory, namely, that the blood in animals (not just men) is moved continuously in a circle by the contraction of the heart.

A careful examination of Harvey's text suggests that Harvey considered the following attributes as especially relevant to the comparison of his theory with Galen's: explanatory power, analogies with accepted theories in other domains, simplicity and logical (internal) consistency. A priori this set of attributes has no more to recommend it than a number of others one might suggest. But for our purposes it is not necessary to reach an agreement on the optimal set of appraising attributes. All that is required here is a generally satisfactory set, or as Herbert Simon would say, a 'satisficing' set.[23] Our primary goal is to show that with respect to each of these attributes, Harvey's theory is superior to Galen's.

a. Explanatory power

Although Galen's theory could not account for any observable phenomena for which Harvey's theory had no explanation, the latter could but the former could not account for the facts that: 1. The amount of blood that passed into the aorta in an hour weighed much more

---

[22] Crombie, *op. cit.*, Vol. II, pp. 226–227.
[23] H. A. Simon and J. G. March, *Organizations*, (New York: John Wiley and Sons, 1958), pp. 140–141.

COST-BENEFIT VERSUS UTILITY ACCEPTANCE RULES 107

than the total amount found in an animal.[24] 2. This amount of blood did not drain all the veins or rupture the arteries.[25] 3. It could not be produced from the juices of ingested aliment or absorbed as nutriment.[26] 4. When the vena cava is closed the heart becomes pale and smaller, and when the aorta is closed the heart becomes deep purple and larger.[27] 5. A middling ligature closing a vein can cause a limb to swell and a tight ligature closing an artery can cause it to turn pale.[28] 6. Dissected bodies have much more blood in their veins than in their arteries and much more in the right ventricle than the left.[29] 7. The valves are situated in order to prevent the passage of blood from the large to the small veins which would cause swelling and rupture.[30] 8. Tumefaction follows a blow to the temple.[31] 9. In phlebotomy a ligature must be applied above the puncture.[32] 10. When a patient undergoing a phlebotomy becomes weaker the blood drains more slowly.[33] 11. A whole system may become contaminated although the originally infected part is apparently sound.[34] 12. Medicine applied externally influences internal organs.[35]

As if all of this were not enough, Harvey includes the following general remark near the end of his penultimate chapter.

> "Finally, reflecting on every part of medicine, physiology, pathology, semeiotics and therapeutics, when I see how many questions can be answered, how many doubts resolved, how much obscurity illustrated by the truth we have declared, ... I see a field of such vast extent in which I might proceed so far, and expatiate so widely, that this my tractate would not only swell out into a volume, ... but my whole life, perchance, would not suffice for its completion."[36]

b. *External analogies*

While the sort of movement envisaged by Galen's theory suggested little more than a perpetually flowing stream, the cyclic motion of Harvey's theory was analogous to and fit together admirably with a

---

[24] Harvey, *loc. cit.*, pp. 74–78, 87.
[25] *Ibid.*, p. 94.
[26] *Ibid.*, pp. 75–78, 80, 87.
[27] *Ibid.*, pp. 46–47, 79.
[28] *Ibid.*, pp. 81, 83.
[29] *Ibid.*, pp. 76, 111.
[30] *Ibid.*, pp. 89–90.
[31] *Ibid.*, p. 85.
[32] *Ibid.*
[33] *Ibid.*, p. 87.
[34] *Ibid.*, p. 96.
[35] *Ibid.*, p. 97.
[36] *Ibid.*, pp. 99–100.

number of ideas and theories that he accepted in other domains. Indeed, immediately following his description of the path followed by the blood, he cites four analogies that would have been familiar to most of his readers.

> "This motion we may be allowed to call circular, in the same way as Aristotle says that the air and the rain emulate [a] the circular motion of the superior bodies; [b] for the moist earth, warmed by the sun, evaporates; the vapours drawn upwards are condensed, and descending in the form of rain, moisten the earth again. [c] By this arrangement are generations of living things produced; [d] and in like manner are tempests and meteors engendered by the circular motion, and by the approach and recession of the sun."[37]

Seven chapters later we find him again arguing in peripatetic fashion that since "under all circumstances" motion generates and preserves "heat and spirits" which are necessary for life, a body must have its "particular seat and fountain, a kind of home and hearth, where ... the original of the native fire, is stored and preserved." This "original" in animals was none other than their pulsating hearts.[38]

c. Simplicity

Harvey's theory was simpler than Galen's in the sense that the former required fewer basic assumptions and ad hoc hypotheses than the latter. In the first place, Galen regarded the liver as the center of the venous system and the heart as the center of the arterial system, with anastomosis between the two systems and "communication between the cavities of the heart" through "the tiny pores which appear above all toward the middle of the partition between the cavities...."[39] Harvey's theory of course had a single center, the heart, and as for "the tiny pores" in the septum, he exclaimed "By Hercules! no such pores can be demonstrated, nor in fact do any such exist."[40] Furthermore, the ad hoc assumption of anything passing through the septum raised more questions than it answered.

> "... how could one of the ventricles extract anything from the other ... when we see that both ventricles contract and dilate simultaneously? Why

---

[37] *Ibid.*, p. 71. Both Crombie (*op. cit.*, Vol. II, pp. 235–237) and W. Pagel, "The position of Harvey and van Helmont in the history of European thought," *Toward Modern Science. Vol. II*, New York: Noonday Press, 1961), pp. 177–182 regard these analogies as highly influential on Harvey's thinking.

[38] Harvey, *loc. cit.*, p. 94; Pagel, *loc. cit.*, p. 181.

[39] Harvey, *loc. cit.*, pp. 34, 38–39; Galen, *On the Natural Faculties*, (London, 1916), translated by A. J. Brock, p. 321.

[40] Harvey, *loc. cit.*, p. 42.

should we not rather believe that the right took spirits from the left, than that the left obtained blood from the right ventricle...? But it is certainly mysterious and incongruous that blood should be supposed to be most commodiously drawn through a set of obscure and invisible ducts, and air through perfectly open passages, at one and the same moment. And why ... is recourse had to secret and invisible porosities, to uncertain and obscure channels, to explain the passage of the blood to the left ventricle, when there is so open a way through the pulmonary veins."[41]

It is perhaps worthwhile to notice here that although the capillaries required by Harvey's theory to permit blood to pass from arteries to veins were as "obscure and invisible" as the "uncertain and obscure channels" through the septum required by Galen's theory,[42] the assumption of the existence of the former did not create more problems than it solved. Indeed, it seems to be primarily this aspect of the hypothesis of invisible capillaries which makes it decidedly not ad hoc.

Second, Galen's view of the *causes* of the movement of "material" to and from the heart was enormously more complicated than Harvey's. The latter's view was mentioned in the fourth point that was cited above following his summary of chapter 14. In Galen's view "almost all parts of the animal" possessed an "attractive faculty" by means of which they obtained their "proper juice," a "retentive faculty" which was responsible for the retention of whatever was of "some benefit," an "alterative faculty" which accounted for the conversion of attracted material into "nourishment," and an "expulsive faculty" that explained the elimination of whatever was not of "some benefit."[43] For example, with respect to the stomach he explains:

"... the attractive faculty in connection with swallowing, the retentive with digestion, the expulsive with vomiting and with the descent of digested food into the small intestine – and digestion itself we have shown to be a process of alteration."[44]

He also distinguishes *"two kinds of attraction*, that by which a vacuum becomes refilled and that caused by appropriateness of quality."[45] All "hollow organs" such as the heart and arteries display both kinds of attraction during diastole, with the former "always attracting lighter matter first" and perhaps from some distance, while the

[41] *Ibid.*
[42] The flow of blood through capillaries was not observed until 1661 by Marcello Malpighi; Crombie, *op. cit.*, Vol. II, p. 234.
[43] Galen, *On the Natural Faculties*, pp. 223–225, 247–249, 307.
[44] *Ibid.*, p. 275.
[45] *Ibid.*, p. 319.

latter "acts frequently ... on what is heavier" and usually nearby.[46]

"The arteries draw into themselves on every side; those arteries which reach the skin draw in the outer air ... those which pass up from the heart into the neck, and that which lies along the spine ... draw mostly from the heart itself; and those which are further from the heart and skin necessarily draw the lightest part of the blood out of the veins."[47]

This should be enough for our purposes. Rather than providing a general causal explanation of the movement of "material" to and from the heart as described in his *On the Functions of Parts of the Human Body* and outlined in Figure 1, Galen has given us reasons to expect *not* that but another kind of movement. For with this explanatory scheme "almost all" of the single arrows in Figure 1 should be replaced by double arrows.[48] Furthermore, considering the fact that he only uses the term 'faculty' "so long as we are ignorant of the true essence of the cause which is operating,"[49] even if his view were internally consistent, it would not be very informative. And finally, to return to our original point, even if the scheme worked, it was much more complicated and contained many more loose ends than Harvey's.

d. Internal consistency

Galen was fully aware that the uncoordinated activity of the "attractive" and "expulsive faculties" could lead to paralysis or chaos. But he thought that the whole system could run smoothly if the faculties operated "consecutively" like inhaling and exhaling.[50] Harvey knew that that explanation was inadequate. If, as noted above, the "ventricles contract and dilate simultaneously" and if they only attract with dilation and repel with contraction, then they could not be exchanging anything "consecutively."[51] So either there was no exchange or the "faculty" scheme was faulty, or, as Harvey claimed, both. Similarly, Harvey claimed that on Galen's view the "spirits"[52] in the aorta

---

[46] *Ibid.*, pp. 317–319, 325.
[47] *Ibid.*, p. 317.
[48] Crombie, *op. cit.*, Vol. I, pp. 164–165 and Fleming, *op. cit.*, completely missed this point, and criticized historians who had referred to a general ebbing and flowing in the whole venous system. See also *d* below.
[49] Galen, *On the Natural Faculties*, p. 17.
[50] *Ibid.*, pp. 303–307.
[51] Harvey, *loc. cit.*, p. 42.
[52] Galen imagined that the vital functions were produced by the activity of three kinds of "spirits," namely, the "vital spirit" of the heart, the "natural spirit" of the liver and the "animal spirit" of the brain. The first "accounted for" the "vital faculty" or "principle of animal life," the second for the "vegetative faculty" or "principle of nutrition and growth" and the third for the "psychic faculty" or "spiritual principle of life." Crombie, *op. cit.*, Vol. I, pp. 163–167.

(which were necessary to the life of the heart as well as every other organ) should have been drawn into the left ventricle as a result of its "attractive faculty," but somehow they always escaped.[53] This is merely a special case of the general point made above, namely, that "almost all" of the arrows on Galen's view in Figure 1 turn out to be both double and single at the same time, which is impossible. Again, Harvey saw that the idea that the mitral valve[54] should allow the "spirituous blood" to pass from the left ventricle to the lungs while at the same time it prevented the "thinner" air from retrogressing through the same channel was plainly inconsistent.[55] And finally, he noted a similar infelicity in the alleged "cooling and cleaning system" operating between the left ventricle and lungs by means of the pulmonary vein. If the mitral valve prevented the "cooling" air from escaping once it arrived in the left ventricle then it could not fail to prevent the "fuliginous vapours" from escaping also, in which case there would be no "cleansing" activity.[56]

This completes my review of Harvey's comparison of his theory with that of Galen's on the movement of the blood and the function of the heart with respect to the four attributes of explanatory power, analogies with other theories, simplicity and internal consistency. The implicit application of CBD with these attributes was *effective*. It produced a decision in favor of Harvey's view over Galen's. Harvey was certainly not "all right," expecially in his selection of acceptable scientific theories in other domains, which is quite understandable.[57] But his errors are only relevant to the content of his argument, not to its logical form. From the logical or methodological point of view, his argument was perfectly nondemonstratively valid. His theory dominated its alternatives and was, therefore, more acceptable. Fortunately (for Harvey at least), unlike most theories, his theory has continued to dominate its alternatives. In fact, by the time he and his contemporaries had passed away, serious alternatives were no longer put forward.[58] Hence, today it is appropriate to describe his

---

[53] Harvey, *loc. cit.*, p. 40.
[54] The mitral valves are located between the auricles and ventricles in the mitral orifaces on both sides of the heart. The one referred to here is on the left side and known as the "bicuspid valve" because it has two flaps or doors.
[55] Harvey, *loc. cit.*, p. 40.
[56] *Ibid.*
[57] According to the first quotation in subsection *b* above, he evidently accepted a geocentric theory of the planetary system.
[58] Graubard, *op. cit.*, pp. 172–176.

theory not merely as more acceptable than its alternatives, but as acceptable.

Considering the results of sections five and six together now, I take it that the effectiveness and efficiency, and therefore, the superiority and preferability of CBD over MEU has been established. Before closing this investigation, however, two more general topics merit our attention. The first pertains to a certain infelicity shared by both MEU and CBD, and will be discussed in the next section. The second concerns various strategies that could be used to increase the effectiveness of CBD and, therefore, strengthen our case for it. These issues are taken up in section eight.

7. AN INFELICITY OF MEU AND CBD

The major drawback of both of these principles is that they do not provide any built-in evaluation for the variety of evidence for or against a hypothesis. If, for example, one hypothesis has a utility of .2 on the basis of a single attribute and a probability of .6 of obtaining its full value, while another hypothesis has a utility value of .2 for each of three attributes (two plus the one on which the other hypothesis is superior) with probabilities of .2 each, then the expected utility of each hypothesis is the same .12. Similarly, neither hypothesis dominates the other. However, the hypothesis with a greater variety of support might plausibly be regarded as warranting a higher assessment. What can be said about this discrepancy?

It seems to me that this problem of assessing variety may be treated in much the same way that voting theorists treat the problem of "no election." That is, we may introduce the attribute of 'variety' into our analysis just as voting theorists introduce the option 'no election' along with the list of candidates.[59] Then, just as a voter is allowed to judge the merits of each candidate in the presence of the option to have the whole election rescinded, our decision-makers are allowed to judge the variety of support for or against a hypothesis and assign it some appropriate value. Whether or not this strategy would work as well for our decision-makers as its analogue works for voters, a priori it certainly seems that it would.

---

[59] See, for example, C. L. Dodgson, "A discussion of the various methods of procedure in conducting elections," reprinted in D. Black, *The Theory of Committees and Elections*, (Cambridge: Cambridge University Press, 1958), pp. 214–222.

8. INCREASING THE EFFECTIVENESS OF CBD

There are a number of ways to increase the effectiveness of CBD, some of which put more severe demands on the number of assumptions and kinds of information required than proponents of CBD would be willing to satisfy. In the remaining paragraphs of this section I shall introduce five general tactics and indicate their peculiar costs.

8.1 You recall that one hypothesis was said to dominate another if *for all* attributes and contingencies the benefits and costs associated with the latter are not preferable to those of the former and for some attribute and contingency the benefits and costs associated with the former are preferable to those of the latter. The italicized phrase 'for all' may be regarded as an abbreviation of the longer locution 'for all $n(n \geq 1)$ relevant attributes and $m(m \geq 1)$ contingencies.' As long as n is sufficiently large, we may weaken the notion of dominance by degrees, by replacing 'for all n' by 'for all $n - 1$,' 'for all $n - 2$,' and so on up to 'for all $n - (n - 1)$.' Thus while strong dominance and dominance place certain requirements on all attributes, $(n - 1)$-dominance puts requirements on all but one attribute, $(n - 2)$-dominance on all but two, etc. For example, a hypothesis whose benefits and costs were preferable to those of another with a single exception in which the latter's benefits (or costs) were preferable to the former might be said to $(n - 1)$-dominate the latter, although it could not dominate the latter. Clearly, the chances of obtaining a single acceptable hypothesis with CBD increase as the degrees of dominance decrease from n. Moreover, no new information is required. On the other hand, it must be assumed that the data from some *prima facie* relevant attribute(s) may be safely ignored. That, of course, may be difficult to justify, especially in the completely general fashion proposed. For notice that, say, $(n - 1)$ -dominance does not specify any particular attribute to be ignored. It merely permits a reversal in any attribute whatsoever, and one may be reluctant to grant such sweeping permission.[60]

Just as one can reduce n to $n - 1$, etc. one can reduce m to $m - 1$, etc. when there are a sufficient number of contingencies. This would produce the same advantages and disadvantages as reductions in n.

Finally, it should be mentioned that one could again use the analogy between this investigation and voting theory, and consider such familiar notions of dominance as simple majority dominance, absolute majority

---

[60] Most of the case histories cited in Bunge, *op. cit.*, seem to have admitted some reversals.

dominance, ⅔ dominance, and so on.⁶¹ For specific n's and m's, all of these phrases could be translated into the 'n minus something' terminology.

8.2 Having considered the apriori elimination of attributes and contingencies, there are two fairly natural tacks to take. One may consider the elimination of specially selected contingencies or of specially selected attributes. I shall discuss the former in this subsection and the latter in the next.

It is a familiar fact that the decision rules known as 'minimax loss,' 'minimax regret,' 'maximin gain,' 'maximax gain' and 'Hurwicz's rule' focus a decision-maker's attention on only some of the contingencies before him. For example, minimax loss tells one to merely review the maximum losses (costs) possible as a result of accepting any hypothesis given each contingency, and to act so as to guarantee the smallest of the maximum losses possible. Similarly, one might eliminate all of the data on benefits from one's analysis and define a concept of minimax-dominance. To determine which hypothesis minimax-dominated which, one would review the maximum costs attached to each hypothesis for every contingency and regard that one as minimax-dominant which insured the smallest maximum possible cost. Concepts of minimax regret-dominance, maximax-dominance, etc. could be constructed analogously.

All of these qualified types of dominance would be easier to obtain than unqualified dominance. So their use would increase the effectiveness of CBD. Furthermore, they do not require any more information. Indeed, some of them require less information, because they completely disregard either benefits or costs. However, this demands the rather bold assumption that such data *and more* can be safely ignored, and a priori there seems to be no justification for this assumption.

8.3 As Miller and others have shown,⁶² there is a general tendency

---

[61] From a formal or logico-mathematical point of view, voting theory and the theory of multi-attribute decision-making are virtually indistinguishable. See, for example, Alex C. Michalos, "A theory of decision-making evaluation," paper read at the Annual Meeting of the Eastern Division of the American Philosophical Association, December 1969; Alex C. Michalos, "Decision-making in committees," *American Philosophical Quarterly*, 7 (1970), pp. 95–116; and Alex C. Michalos, "The impossibility of an ordinal measure of acceptability," *The Philosophical Forum*, to be published.

[62] G. A. Miller, "The magical number seven, plus or minus two," *Psychological Review*, 63 (1956), pp. 81–97; J. R. Hayes, "Human data processing limits in decision-making," *Electronics System Division Report* EDS-TDR-62-48, 1962; and R. N. Shepard, "On subjectively optimum selection among multi-attribute alternatives," *Human Judgments and Optimality*, ed. M. W. Shelly and G. L. Bryan, (New York: John Wiley and Sons, 1964), pp. 257–281.

for decision-makers to unwittingly let one or two of many relevant attributes determine their final judgment. Moreover, as MacCrimmon and Raiffa[63] have recently emphasized, for one reason or another, a decisionmaker may *choose* to regard one attribute as more important than all of the others together. In the latter case then, concepts of specific attribute dominance might be defined such as 'probability-dominance,' 'explanatory power-dominance,' etc. to be used with CBD. While such an approach presupposes interattribute comparisons of importance and, therefore, additional evaluative criteria, "weighing" devices, assumptions and justifications, it is still less demanding than MEU. Hence, with this modification, CBD would probably (depending primarily on the number of vitally important attributes selected) still be more effective than MEU.

8.4 If the total expulsion of some attributes and/or contingencies from the set of relevant evidence seems unjustifiable, a less drastic procedure may seem attractive. It has already been noted that evaluative criteria and "weighing" devices would have to be developed in order to select the one or two supremely important attributes mentioned in the previous paragraph. *If* every attribute could be assigned a numerical value indicating its weight of importance *and* all attribute values could also be expressed numerically, then each hypothesis could be assigned a numerical value equal to the sum of its weighted attribute values and the hypothesis with the largest sum could be regarded as the most acceptable. By requiring all weights of importance to be real numbers greater than zero, one could be certain that every relevant attribute had *some* influence on the total evaluation sum for each hypothesis. Neat as it sounds, such a procedure could not be practicable, because it demands even more information and assumptions, and could not be more effective than MEU.

8.5 As Simon[64] and Ellsberg[65] have insisted, it is sometimes easier to determine that something is unsatisfactory than it is to determine just how satisfactory something is. It is usually easier to decide which shirts, suits, socks, or ties "just won't do" than it is to decide which of a couple fairly decent ones one should buy. Following Simon, we may say that an attribute value is satisficing if and only if hypotheses with such a value could in every sense of this term be acceptable. It follows, then, that any hypothesis with a non-satisficing value for some at-

---

[63] MacCrimmon, *op. cit.*, and Raiffa, *op. cit.*,
[64] Simon and March, *op. cit.*
[65] Ellsberg, *op. cit.*

tribute in some contingency cannot be acceptable. Hence, such hypotheses may be immediately eliminated from consideration, with CBD applied to the remainder. By providing a good reason for rejecting some hypotheses that might otherwise remain in the set of live options, a review of attribute values from a satisficing point of view could increase the a priori chance of obtaining a single acceptable hypothesis with CBD. The apparent additional information required is the minimum satisficing attribute value for every attribute. Even if such comprehensive data was not available, however, it might still be worthwhile (i.e., increase the effectiveness of CBD) to know *some* minimum satisficing values. Naturally the identification of such valeus presupposes assumptions and justifications for the evaluative criteria employed.

8.6 Finally, it should be noted that one could combine some of the tactics described in 8.1–8.5[66] For examples, one could use a satisficing review of attribute values (8.5) along with a weaker concept of dominance (8.1); a contingency eliminating rule (8.2) with a weaker concept of dominance; (8.1), (8.2) and (8.5) together; and so on. What must always be remembered, of course, is that increases in the number of tactics employed create corresponding increases in the number of assumptions and justifications required. Furthermore, the very reason CBD has been recommended here in the first place is that it is supposed to reduce the latter without excessive costs.

9. CONCLUSION

Since I have summarized the argument for my central thesis at the end of section four, is not necessary to repeat it here. All that remains to be said now is that what is required at this point is a strong defense of a particular set of relevant attributes or, in Bunge's words, "assaying criteria" for the evaluation of all hypotheses. Perhaps no single set will do for all kinds of hypotheses. What we require of acceptable laws may be different from what we require of acceptable theories. Attributes that are relevant for the determination of the acceptability of ordinary sentences (rather than laws or theories) may well be something else again. At any rate, if my case for CBD over MEU has been argued persuasively, then it is clearly the relevant attributes or "assaying criteria" that should be the focus of our attention now.

---

[66] This is also suggested by MacCrimmon, *op. cit.*

LIST OF REFERENCES

BOOKS

Ackoff, R. L. *Scientific Method.* New York: John Wiley and Sons, 1962.
Arrow, K. L. *Social Choice and Individual Values.* New York: John Wiley and Sons, 1951.
Barker, S. F. *Induction and Hypothesis: A Study of the Logic of Confirmation.* Ithaca, New York: Cornell University Press, 1962.
Buchdahl, G. *Metaphysics and the Philosophy of Science.* Oxford: B. H. Blackwell, 1969.
Carnap, R. *The Logical Foundations of Probability.* Chicago: University of Chicago Press, 1950.
— *Continuum of Inductive Methods.* Chicago: University of Chicago Press, 1952.
Crombie, A.C. *Medieval and Early Modern Science, Vol. I and II.* Garden City: Doubleday, 1959.
Ellsberg, D. *Risk, Ambiguity, and the Savage Axioms.* P-2173, The RAND Corporation, Santa Monica, 1961.
Fishburn, P. C. *Decision and Value Theory.* New York: John Wiley and Sons, 1964.
Fisher, R. A. *Statistical Methods and Scientific Inference.* 2d. ed. London: Oliver and Boyd, 1959.
Galen. *On the Natural Faculties.* translated by A. J. Brock, London, 1916.
Gnedenko, B. V., and A. Y. Khinchin. *An Elementary Introduction to the Theory of Probability.* San Francisco: W. H. Freeman and Co., 1961.
Graubard, M. *Circulation and Respiration,* New York: Harcourt, Brace and World, 1964.
Hanson, N. R. *Patterns of Discovery.* Cambridge: The University Press, 1958.
Harvey, W. *An Anatomical Disquisition on the Motion of the Heart and Blood in Animals;* Willis's translation revised and edited by A. Bowie, London, 1889 and reprinted in *Classics of Medicine and Surgery,* ed. C. M. B. Camac, New York: Dover Pub., 1959.
Hintikka, J. and Suppes, P. *Aspects of Inductive Logic.* Amsterdam: North-Holland Pub. Co., 1966.
Jeffreys, H. *Theory of Probability.* Oxford: Clarenden Press, 1939.
Jevons, W. *The Principles of Science.* New York: Dover Pub. Inc., 1958.
Keynes, J. *A Treatise on Probability.* London: Macmillan and Co., 1921.
Knight, F. H. *On the History and Method of Economics.* Chicago: University of Chicago Press, 1956.
Kneale, W. *Probability and Induction.* Oxford; Oxford University Press, 1949.

Kuhn, T. S. *The Structure of Scientific Revolutions.* Chicago: University of Chicago Press, 1962.
Laplace, P. S. de. *Theorie Analytique des Probabilites.* Paris: 1812.
Levi, I. *Gambling with Truth.* New York: Alfred A. Knopf, 1967.
Little, I. M. D. *A Critique of Welfare Economics.* New York: Oxford University Press, 1950.
Margenau, H. *The Nature of Physical Reality.* New York: McGraw-Hill, 1950.
Mehlberg, H. K. *The Reach of Science.* Toronto: University of Toronto Press, 1958.
Michalos, A. C. *Principles of Logic.* Englewood Cliffs: Prentice-Hall, Inc., 1969.
— *Foundations of Decision-Making.* New York: American Elsevier Pub. Co. Inc., in preparation.
Pap, Arthur. *An Introduction to the Philosophy of Science.* New York: The Free Press, 1962.
Popper, K. R. *The Logic of Scientific Discovery.* New York: Science Editions, Inc., 1961.
— *Conjectures and Refutations.* New York: Basic Books, 1962.
Reichenbach, Hans. *Theory of Probability.* 2d ed. Berkeley, California: University of California Press, 1949.
— *The Rise of Scientific Philosophy.* Berkeley, California: University of California Press, 1959.
Rescher, N. *Introduction to Value Theory.* Englewood Cliffs: Prentice-Hall, Inc., 1969.
Rothenberg, J. *The Measurement of Social Welfare.* Englewood Cliffs: Prentice-Hall, Inc., 1961.
Salmon, W. C. *The Foundations of Scientific Inference.* Pittsburgh: University of Pittsburgh Press, 1966.
Savage, L. J., et al. *The Foundations of Statistical Inference.* New York: John Wiley and Sons, 1962.
Scheffler, I. *The Anatomy of Inquiry.* New York: Alfred A. Knopf, 1963.
Simon, H. A. and March, J. G. *Organizations.* New York: John Wiley and Sons, 1958.
Singer, C. *A Short History of Anatomy and Physiology from the Greeks to Harvey.* New York: Dover Pub., 1957.
Toulmin, S. *The Philosophy of Science.* New York: Harper and Brothers, 1960.
von Mises, R. *Probability, Statistics and Truth.* 2d ed. rev. translated by Hilda Geiringer. London: George Allen and Unwin, 1957.
von Neumann, J. and Morgenstern, O. *The Theory of Games and Economic Behavior.* Princeton: Princeton University Press, 1947.
Williams, D. C. *The Ground of Induction.* Cambridge, Mass., 1947.

ARTICLES

Agassi, J. "Corroboration versus induction," *B.J.P.S.*, IX (1958), pp. 311–317.
Alexander, H. G. "The paradoxes of confirmation," *B.J.P.S.*, IX (1958), pp. 227–233.
— "The paradoxes of confirmation: a reply to Dr. Agassi," *B.J.P.S.*, X (1959), pp. 229–234.
Ayer, A. J. "The conception of probability as a logical relation," *Observation and Interpretation in the Philosophy of Physics*, ed. S. Körner, New York: Dover Pub., 1957, pp. 12–17.
Bar-Hillel, Y. "Comments on 'Degree of Confirmation' by Professor K. R. Popper," *B.J.P.S.*, VI (1955), pp. 155–157.

LIST OF REFERENCES

— "Further comments on probability and confirmation, a rejoinder to Professor Popper," *B.J.P.S.* VII (1956), pp. 245–248.
— "On an alleged contradiction in Carnap's theory of inductive logic," *Mind*, LXXIII (1964), pp. 265–267.
Bunge, M. "The weight of simplicity in the construction and assaying of scientific theories," *Philosophy of Science*, 28, (1961), pp. 120–149.
Burks, A. W. "The pragmatic-Humean theory of probability and Lewis' theory," *The Philosophy of C. I. Lewis*, ed. P. A. Schilpp, LaSalle, Illinois: Open Court Press, 1968, pp. 415–464.
Carnap, Rudolf. "Testability and meaning," *Philosophy of Science*, III (1936), pp 419–471 and IV (1937), pp. 1–40..
— "Truth and confirmation," *Readings in Philosophical Analysis*. ed. H. Feigl and W. Sellars, New York: Appleton-Century-Crofts, Inc., 1949, pp. 119–127.
— "The two concepts of probability," *Readings in the Philosophy of Science*. ed. H. Feigl and M. Brodbeck, New York: Appleton-Century-Crofts, Inc., 1953, pp. 438–455.
— "On inductive logic," *Philosophy of Science*, XII (1945), pp. 72–97.
— "Statistical and inductive probability," *The Structure of Scientific Thought*. ed. E. H. Madden, Boston: Houghton Mifflin, Co., 1960, pp. 269–278.
— "Remarks on Popper's note on content and degree of confirmation," *B.J.P.S.* VII (1956), pp. 244–245.
— "The aim of inductive logic," *Logic, Methodology and Philosophy of Science*. ed. E. Nagel, P. Suppes and A. Tarski, Stanford, California: Stanford University Press, 1962, pp. 303–318.
— "Replies and systematic expositions," *The Philosophy of Rudolf Carnap*, ed. P. A. Schilpp, LaSalle, Illinois: Open Court Press, 1963, pp. 859–1013.
— "Remarks on probability," *Philosophical Studies*, XIV (1963), pp. 63–75.
Chisholm, R. M. "Lewis' ethics of belief." *The Philosophy of C. I. Lewis*, ed. P. A. Schilpp, LaSalle, Illinois: Open Court Press, 1968, pp. 223–242.
Dodgson, C. L. (Lewis Carroll), "A discussion of the various methods of procedure in conducting elections," reprinted in Black, D. *The Theory of Committees and Elections*, Cambridge: Cambridge University Press, 1963, pp. 214–222.
Ellsberg, D. *Risk, Ambiguity, and the Savage Axioms*, P-2173, The RAND Corporation, 1961.
Fleming, D. "Galen on the motions of the blood in the heart and lungs," *Isis*, 46 (1955), pp. 14–21.
Feyerabend, P. K. "How to be a good empiricist – A plea for tolerance in matters epistomological," *Philosophy of Science: The Delaware Seminar II*, ed. B. Baumrin, New York: John Wiley and Sons, Inc., 1963, pp. 3–40.
Good, I. J. "Corroboration, explanation, evolving probability, simplicity and a sharpened razor," *B.J.P.S.*, 19, (1968), pp. 123–143.
Grunbaum, Adolf. "The falsifiability of theories: total or partial?" *Boston Studies in the Philosophy of Science*. ed. M. W. Wartofsky, Dordrecht, Holland: D. Reidel Pub. Co., 1963, pp. 178–195.
Hanson, N. R. "Is there a logic of discovery?" *Current Issues in the Philosophy of Science*, ed. H. Feigl and G. Maxwell, New York: Holt, Reinhart and Winston, Inc., 1961, pp. 20–34.
Harsanyi, J. C. "Cardinal Welfare, individualistic ethics and interpersonal comparisons of utility," *Journal of Political Economy*, 63 (1955), pp. 309–321.
Hayes, J. R. "Human data processing limits in decision making," *Electronics System Division Report*, ESD-TDR-62-48, 1962.
Helmer, O. and P. Oppenheim. "A syntactical definition of probability and of degree of confirmation," *Journal of Symbolic Logic*, X (1945), pp. 25–60.

LIST OF REFERENCES

Hempel, C. G. "A purely syntactical definition of confirmation," *Journal of Symbolic Logic*, VIII (1943), pp. 122–143.
— "Studies in the logic of confirmation," *Mind*, LIV (1945), pp. 1–26, 97–121.
— and P. Oppenheim, "A definition of 'Degree of Confirmation'," *Philosophy of Science*, XII (1945), pp. 98–115.
— "Problems and changes in the empiricist criterion of meaning," *Revue Internationale de Philosophie*, II (1950), pp. 41–63.
— "Fundamentals of concept formation in empirical science," *International Encyclopedia of Unified Science*, Vol. II (1960).
— "Deductive-nomological vs. statistical explanation," *Minnesota Studies in the Philosophy of Science*. Vol. III ed. H. Feigl and G. Maxwell, Minneapolis: University of Minnesota Press, 1962, pp. 98–169.
— "Inductive inconsistencies," *Logic and Language: Studies Dedicated to Professor Rudolf Carnap on the Occasion of His 70th Birthday*, Dordrecht, Holland: D. Reidel Pub. Co., 1962, pp. 128–139.
— "Recent problems of induction," *Mind and Cosmos*, ed. R. G. Colodny, Pittsburgh: University of Pittsburgh Press, 1966, pp. 112–134.
Hildreth, C. "Alternative conditions for social orderings," *Econometrica*, 21 (1953), pp. 81–91.
Hilpinen, R. "Rules of acceptance and inductive logic," *Acta Philosophica Fennica*, Fasc. XXII (1968), pp. 1–134.
Hintikka, J. "Towards a theory of inductive generalization," *Proceedings of the 1964 Congress for Logic, Methodology and Philosophy of Science*, ed. Y. Bar-Hillel, Amsterdam: North-Holland Pub. Co., 1965, pp. 274–288.
— "A two-dimensional continuum of inductive methods," *Aspects of Inductive Logic*, ed. J. Hintikka and P. Suppes, Amsterdam: North-Holland Pub. Co., 1966, pp. 113–132.
— "Induction by enumeration and induction by elimination," *The Problem of Inductive Logic*, ed. I. Lakatos, Amsterdam: North-Holland Pub. Co., 1968, pp. 191–216.
Jeffrey, R. C. "Popper on the rule of succession," *Mind*, LXXIII (1964), p. 129.
Kemeny, J. G. "Review of K. R. Popper 'Degree of Confirmation'," *Journal of Symbolic Logic*, XX (1955), pp. 304–305.
Lehrer, K. "Knowledge and probability," *Journal of Philosophy*, LXI (1964), pp. 368–372.
— "Descriptive completeness and inductive methods," *The Journal of Symbolic Logic*, XXVIII (1963), pp. 157–160.
Lakatos, I. "Changes in the problem of inductive logic," *The Problem of Inductive Logic*, ed. I Lakatos, Amsterdam: North-Holland Pub. Co., 1968, pp. 315–417.
Laudan, L. "Theories of scientific method from Plato to Mach," *History of Science*, 6 (1968), pp. 1–63.
Levi, I. "On the seriousness of mistakes," *Philosophy of Science*, 29 (1962), pp. 47–65.
— "Deductive cogency in inductive inference," *Journal of Philosophy*, LXII (1965), pp. 68–77.
MacCrimmon, K. R. *Decision-making Among Multiple-Attribute Alternatives: A Survey and Consolidated Approach*, Memorandum RM–4823–ARPA, The RAND Corporation, Santa Monica, 1968.
Mackie, J. L. "The paradox of confirmation," *B.J.P.S.*, XIII (1963), pp. 265–277.
Michalos, A. C. "Two theorems of degree of confirmation," *Ratio*, 7 (1965), pp. 196–198.

- "Estimated utility and corroboration," *B.J.P.S.*, 16 (1966), pp. 327–331.
- "Postulates of rational preference," *Philosophy of Science*, 34 (1967), pp. 18–22.
- "Descriptive completeness and linguistic variance," *Dialogue*, 6 (1967), pp. 224–228.
- "An alleged condition of evidential support," *Mind*, 78 (1969), pp. 440–441.
- "A Theory of decision-making evaluation," a paper read at the Annual Meeting of the Eastern Division of the American Philosophical Association, 1969.
- "Analytic and other 'dumb' guides of life," *Analysis*, 30 (1970), pp. 121–123.
- "Decision-making in committees," *American Philosophical Quarterly*, 7 (1970), pp. 95–116.
- "Positivism versus the hermeneutic-dialectic school," *Theoria*, 35 (1969), pp. 267–278.
- "The costs of decision-making," *Public Choice*, 9, (1970).
- "Cost-benefit versus expected utility acceptance rules," *Theory and Decision*, 1 (1970).
- "Rules of acceptance and inductive logic: a critical review," *Philosophy of Science*, to be published.
- "The impossibility of an ordinal measure of acceptability," *The Philosophical Forum*, to be published.
- "Efficiency and morality," Paper read at the Annual Meeting of the Western Division of the American Philosophical Association, May 1970.

Miller, G. A. "The magical number seven, plus or minus two," *Psychological Review*, 63 (1956), pp. 81–97.

Milnor, J. "Games against nature," *Decision Processes*, ed. R. M. Thrall, C. H. Coombs and R. L. Davis, New York: John Wiley and Sons, 1960, pp. 49–60.

Pagel, W. "The position of Harvey and van Helmont in the history of European thought," *Toward Modern Science Vol. II*, ed. R. M. Palter, New York: Noonday Press, 1961, pp. 175–191.

Popper, K. R. "Degree of confirmation," *B.J.P.S.*, V (1954), pp. 143–149.
- "'Content' and 'Degree of Confirmation': A reply to Dr. Bar-Hillel," *B.J.P.S.* VI (1955), pp. 157–163.
- "Adequacy and consistency: A second reply to Dr. Bar-Hillel," *B.J.P.S.*, VII (1956), pp. 249–256.
- "A third note on degree of corroboration or confirmation," *B.J.P.S.*, VIII (1958), pp. 394–403.
- "On Carnap's version of Laplace's rule of succession," *Mind*, LXXI (1962), pp. 69–73.

Prest, A. R. and Turvey, R. "Cost-benefit analysis: a survey," *The Economic Journal*, 75 (1965), pp. 683–735.

Pruzan, P. M. "Is cost-benefit analysis consistent with the maximization of expected utility?" *Operational Research and the Social Sciences*, ed. J. R. Lawrence, London: Tavistock Pub., 1966, pp. 319–336.

Salmon, W. C. "Vindication of induction," *Current Issues in the Philosophy of Science*, ed. H. Feigl and G. Maxwell, New York: Holt, Rinehart and Winston, 1961, pp. 254–256.
- "On vindicating induction," *Philosophy of Science*, XXX (1963), pp. 252–261.
- "Inductive inference," *Philosophy of Science*, Vol. II of *The Delaware Seminar* ed. B. Baumrin, New York: John Wiley and Sons, Inc., 1963, pp. 341–370.

Scriven, Michael. "The key property of physical laws – inaccuracy," *Current Issues in the Philosophy of Science*, ed. H. Feigl and G. Maxwell, New York: Holt, Rinehart and Winston, 1961, pp. 91–101.

Shepard, R. N. "On subjectively optimum selection among multi-attribute alternatives," *Human Judgments and Optimality*, ed. M. W. Shelly and G. L. Bryan, New York: John Wiley and Sons, 1964, pp. 257–281.

Sleigh, R. C., Jr. "A note on some epistemic principles of Chisholm and Martin," *Journal of Philosophy*, LXI (1964), pp. 216–218.

Stedry, A. C. and Charnes, A. "The attainment of organization goals through appropriate selection of subunit goals," *Operational Research and the Social Sciences*, ed. J. R. Lawrence, London: Tavistock Pub., 1966, pp. 147–164.

Vincent, R. H. "A note on some quantitative theories of confirmation," *Philosophical Studies*, XII (1961), pp. 91–92.

— "On my cognitive sensibility," *Philosophical Studies*, XIV (1963), pp. 77–79.

— "Concerning an alleged contradiction," *Philosophy of Science*, XXX (1963), pp. 189–194.

Watkins, J. W. N. ' Between analytic and empirical," *Philosophy*, XXXII (1957), pp. 112–131.

— "A rejoinder to Professor Hempel's reply," *Philosophy*, XXXIII (1958), pp. 349–355.

— "When are statements empirical?" *B.J.P.S.*, XI (1960), pp. 287–308.

— "Confirmation without background knowledge," *B.J.P.S.*, XI (1960), pp. 318–320.

Whitely, C. H. "Hempel's paradoxes of confirmation," *Mind*, LIV (1945), pp. 156–158.

Wilkie, J. S. "Harvey's immediate debt to Aristotle and to Galen," *History of Science*, 4, (1965), pp. 103–124.

Will, F. L. "The preferability of probable beliefs," *Journal of Philosophy*, LXII (1965), pp. 57–67.

Williams, D. C. "Professor Carnap's philosophy of probability," *Philosophy and Phenomenological Research*, XIII (1952), pp. 103–121.

Williams, P. M. "The structure of acceptance and its evidential basis," *B.J.P.S.*, 19 (1969), pp. 325–344.

Wilson, C. Z. and Alexis, M. "Basic frameworks for decisions," *Journal of the Academy of Management*, 5 (1962), pp. 151–164.

INDEX

Acceptability, 3–4
Acceptance rule AG_n, 87–90
Acceptance rule AS, 90–91
Ackoff, R. L., 99
Agassi, J., 55
Alexander, H. G., 55
Amalgamation procedures, 103
Arrow, K. J., 97
Attributive constituents, 83
Ayer, A. J., 79
Bar-Hillel, Y., 1, 27, 34–48, 62
Barker, S. F., 99
Bayesian conditionalization, 74
Black, M., 41
Bryan, G. L., 114
Broad, C. D., 71
Buchdahl, G., 96
Bunge, M., 96
Burks, A. W., 99
Calculus of probability, 19
Carnap, R.,
    acceptance, 5
    acceptability and probability, 4–8
    analytic guides, 79
    best hypothesis, 7
    explicandum, 1
    explicatum, 1
    inductive probability, 6
    language families, 42
    logic of discovery, 74–79
    restricted simple laws, 40
    standing up to tests, 5
    tenth adequacy condition, 57–59
    testworthy theories, 4
    theory appraisal, 75
    three concepts of confirmation, 28
    unrestricted simple laws, 40
    well-established theories, 6
CBD, 94–96

Chisholm, R. M., 94
CLI and H–H system, 89
Closure, 87
Cohen, L. J., 90
Constituents, 84
Coombs, C. H., 99
Cost-benefit dominance, 94–96
Criterion of linguistic invariance, 64
Crombie, A. C., 104, 108, 110
Ct-predicates, 83
Davis, R. L., 99
Degree of confirmation,
    and probability, 16–33
    singular sentences, 60–64
    unrestricted universals, 34–59
Difference in probability, 31
Dominance concepts, 113–116
Dumb guides, 80–82
Effectiveness, 97
Efficiency, 97
Efficient growth, 15
Efficient inquiry, 14
Ellsberg, D., 115
Epistemic utility, 100
Fact correcting theories, 77
Feigl, H., 10
Feyerabend, P. K., 68, 79
Fishburn, P. C., 99
Fleming, D., 104
Galen, 104–112
General multiplication rule, 19
Good, I. J., 96
Graubard, M., 104
Grunbaum, A., 41
Guides of life, 79–82
Hanson, N. R., 68
Harsanyi, J., 97, 103
Harvey, W., 104–112
Hayes, J. R., 114

INDEX

Helmer, O., 21
Hempel, C. G., 12, 39, 64, 100, 101
Hildreth, C., 97, 103
Hilpinen, R., 83–91
Hintikka, J., 2, 44, 83–91
H–H system, 83–91
Instance confirmation, 49–59
Jeffrey, R. C., 62
Jeffreys, H., 71
Kemeny, J. G., 16, 26, 29
Keynes, J. M., 1, 71
Kneale, W., 11, 79
Kuhn, T., 78
Lakatos, I., 2, 3, 70–83
Laplace's rule of succession, 61
Laudan, L., 96
Lawrence, J. R., 93
Lehrer, K., 64–69
Levi, I., 94, 100, 101
Linguistic preferences, 68–69
Logical width, 42–43
Lottery paradox, 87
MacCrimmon, K. R., 92, 115, 116
Mackie, J. L., 55
Malcolm, N., 41
March, J. G., 106
Margenau, H., 96
Maximax gain, 114
Maxwell, G., 10
Mehlberg, H. K., 41
MEU, 93–94
Miller, G. A., 114
Milnor, J., 99
Minimax, 114
Morgenstern, O., 100
von Neumann, J., 99
Non-negative value rule, 19
Normal science, 78
Numerically universal statement, 40
Oppenheim, P., 21
Pagel, W., 108
Paradox of confirmation, 55–57
Pareto, V., 97
Performs better, 97
Popper, K. R.,
    analytic guides, 79
    degree of confirmability, 4
    degree of corroboration, 5
    description argument, 9–15
    empirical content, 8
    non-existence statements, 40
    paradoxes of confirmation, 53–54
    probability and confirmation, 19–21, 23–25
    singular predictive inference, 62
    three concepts of confirmation, 27–28
    verisimilitude, 76
Popper-Carnap dispute, 1–2
Pragmatic utility, 100
Preferability, 97
Prest, A. R., 93
Prudential growth, 15
Prudential inquiry, 14
Pruzan, P. M., 92
Q-predicate, 41–43
Q-predicate expression, 42
Qualified instance confirmation, 49–59
Raiffa, H., 93
Ramsey-DeFinetti theorem, 75
Reichenbach, H., 1, 71
Rescher, N., 103
RGC, 65
Rothenberg, J., 97
Rule of greater completeness, 65
Rule of maximizing estimated utility, 4
Salmon, W. C.,
    analytic guides, 79
    C-functions, 66–69
    linguistic invariance, 64
Scheffler, I., 55
Scriven, M., 10
Shepard, R. N., 114
Shelly, M. W., 114
Simon, H. A., 106, 115
Singer, C., 104
Singular predictive inference, 60–69
Special addition rule, 19
Standard-gamble, 99–101
Strictly universal statement, 40
Summation rule, 19
Suppes, P., 83
Thrall, R. M., 99
Toulmin, S., 10
Turvey, R., 93
Unrestricted universals, 40
Verification, 41
Vincent, R. H., 21
Wartofsky, M. W., 41
Watkins, J. W. N., 55